ION EXCHANGE AND SOLVENT EXTRACTION

ION EXCHANGE AND SOLVENT EXTRACTION.

A SERIES OF ADVANCES

Volume 7

EDITED BY

JACOB A. MARINSKY
Department of Chemistry
State University of
New York at Buffalo
Buffalo, New York

YIZHAK MARCUS
Department of Inorganic Chemistry
The Hebrew University
Jerusalem, Israel

MARCEL DEKKER, INC.   New York and Basel

COPYRIGHT © 1977 BY MARCEL DEKKER, INC. ALL RIGHTS RESERVED.

Neither this book nor any part may be reproduced or transmitted in any form or by any means, electronic or mechanical, including photocopying, microfilming, and recording, or by any information storage and retrieval system, without permission in writing from the publisher.

MARCEL DEKKER, INC.
270 Madison Avenue, New York, New York 10016

LIBRARY OF CONGRESS CATALOG CARD NUMBER: 66-29027

ISBN: 0-8427-6571-0

Current printing (last digit):
10 9 8 7 6 5 4 3 2 1

PRINTED IN THE UNITED STATES OF AMERICA

PREFACE

In an earlier volume of this series (Volume 2) a review of developments leading to the then current state of ion-exchange resin synthesis was presented to the reader. In the course of its presentation the general aspects of resin preparation by electrophilic substitution of crosslinked polystyrenes were discussed. Steps taken to meet performance requirements for the increasing spectrum of their specialized applications were emphasized.

The first two chapters of this volume are addressed to these topics once again. In the first chapter, kinetic models for heterogeneous reaction of electrophilic reactants in solid-fluid systems are developed to provide a better understanding of the rate processes involved in the interphase mass transfer phenomena. Better synthetic procedures and improved resin products are predicted to be accessible with proper application of this knowledge. In the second chapter the preparation of different polymeric structures to develop ion-exchange resins with special performance characteristics and properties for particular applications is detailed.

The sizable progress in theoretical and practical aspects of ion-exchange resin synthesis that has been made since discussion of these topics earlier in this series has made return to these subjects most appropriate.

The employment of spectral techniques to study the structure and properties of zeolites has been carefully examined in Chapter 3. Their use to identify and examine the ion species sorbed by zeolites and organic ion exchangers has been evaluated as well. Site definition, site geometry and the type of ion binding can be inferred from such identification and examination of ion species. The spectral techniques, by contributing to identification of complex species sorbed, have led to a better understanding of interactions at ion-exchange sites.

The properties of natural ion-exchange materials with emphasis on clay minerals, are examined in Chapter 4. Their influence on the composition of natural water is undisputed and comprehension of the ion-exchange processes operative in these materials has been sought to facilitate, at least qualitatively, anticipation of the fate of various dissolved substances in natural waters. For this purpose the suitability of various physicochemical descriptions of ion-exchange in natural waters has been evaluated.

Finally, in Chapter 5, the invention of ion-exchange resins with equilibrium properties that show a considerable variation with temperature is described. With these resins hot water can be used in place of chemicals for the regeneration of salt-loaded forms to their acid and base forms. In the course of this presentation the fundamental physical-chemical factors important to development of the optimum product is very aptly demonstrated.

The intent of this series, to examine the ion-exchange phenomenon in a fundamental way continues to be well served in this volume by Chapters 1, 2 and 3. The importance of ion-exchange phenomena in natural processes has not previously been considered in this series and Chapter 4, which deals with the properties of natural ion-exchange materials, is the first of a series of ventures in this direction that are contemplated for future volumes. The complete analysis of the development of a unique ion-exchange product for an important and useful application that is presented in the final chapter was included in this volume because it provides a classic example of scholarly research and development in the ion-exchange field.

As in the earlier volumes of this series, emphasis on extension of the boundaries of a problem under discussion was sought from the authors of this volume who were selected on the basis of

PREFACE

their special knowledge in a particular area. It is believed that each chapter, through such emphasis, is provided with a broader base than a technical paper and more information than a review paper, a continuing goal of this series.

Jacob A. Marinsky

CONTRIBUTORS TO VOLUME 7

B. A. Bolto, CSIRO, Division of Chemical Technology, South Melbourne, Australia

V. A. Davankov, Institute of Organo-Element Compounds, Academy of Sciences, Moscow, U. S. S. R.

Shimon Goldstein, Department of Chemistry, Technion-Israel Institute of Technology, Haifa, Israel

Carla Heitner-Wirguin, Department of Inorganic and Analytical Chemistry, Hebrew University, Jerusalem, Israel

Michael M. Reddy, Division of Laboratories and Research, New York State Department of Health, Albany, New York

S. V. Rogozhin, Institute of Organo-Element Compounds, Academy of Sciences, Moscow, U. S. S. R.

Gabriella Schmuckler, Department of Chemistry, Technion-Israel Institute of Technology, Haifa, Israel

M. P. Tsyurupa, Institute of Organo-Element Compounds, Academy of Sciences, Moscow, U. S. S. R.

D. E. Weiss, CSIRO, Division of Chemical Technology, South Melbourne, Australia

CONTENTS

Preface   iii

Contributors to Volume 7   vii

Contents of Other Volumes   xi

Chapter 1: INTERPHASE MASS TRANSFER RATES OF CHEMICAL REACTIONS WITH CROSSLINKED POLYSTYRENE   1

Gabriella Schmuckler and Shimon Goldstein

I.	Introduction	1
II.	Classification According to Mechanism	2
III.	Rate Equations for Shell-Progressive Reactions	4
IV.	Chloromethylation--A Homogeneous Reaction Model	19
V.	Application of the Shell-Progressive Model to Ion-Exchange Processes	25
	Acknowledgment	27
	References	27

Chapter 2: INFLUENCE OF POLYMERIC MATRIX STRUCTURE ON PERFORMANCE OF ION-EXCHANGE RESINS   29

V. A. Davankov, S. V. Rogozhin, and M. P. Tsyurupa

I.	Introduction	29
II.	The Preparation of Different Polymeric Matrix Structures and Their Properties	32
III.	Conclusion	73
	References	74

Chapter 3: SPECTROSCOPIC STUDIES OF ION EXCHANGERS   83

Carla Heitner-Wirguin

I.	Introduction	84
II.	Preparation of Samples for Spectral Measurements and the Evaluation of Quantitative Parameters	84
III.	Spectroscopic Studies on Zeolites	89
IV.	Organic Ion Exchangers	116
V.	Addendum	154
	References	157

Chapter 4:		ION-EXCHANGE MATERIALS IN NATURAL WATER SYSTEMS	165
		Michael M. Reddy	
	I.	Introduction	166
	II.	Ion-Exchange Reactions in Natural Waters	168
	III.	Relationship Between Clay Mineral Structure and Ion-Exchange Properties	175
	IV.	Ion-Exchange Equilibria and Selectivity Expressions for Components of Natural Exchange Systems	179
	V.	Experimental Measurements of Ion-Exchange Selectivity for Components of Natural Water Systems	193
	VI.	Summary	215
		References	215
Chapter 5:		THE THERMAL REGENERATION OF ION-EXCHANGE RESINS	221
		B. A. Bolto and D. E. Weiss	
	I.	Introduction	222
	II.	Resin Equilibria	224
	III.	Resin Kinetics	253
	IV.	Operation of the Process	269
	V.	Engineering Aspects	274
	VI.	Economic Considerations	281
	VII.	Future Developments	285
		Acknowledgments	286
		References	286
Index			291

CONTENTS OF OTHER VOLUMES

Volume 1

TRANSPORT PROCESSES IN MEMBRANES, S. *Roy Caplan and Donald C. Mikulecky*, Polymer Department, Weizmann Institute of Science, Rehovoth, Israel

ION-EXCHANGE KINETICS, *F. Helfferich*, Shell Development Company, Emeryville, California

ION-EXCHANGE STUDIES OF COMPLEX FORMATION, *Y. Marcus*, Department of Inorganic Chemistry, Hebrew University, Jerusalem, Israel

LIQUID ION EXCHANGERS, *Erik Högfeldt*, Department of Inorganic Chemistry, Royal Institute of Technology, Stockholm, Sweden

PRECISE STUDIES OF ION-EXCHANGE SYSTEMS USING MICROSCOPY, *David H. Freeman**, National Bureau of Standards, Washington, D. C.

HETEROGENEITY AND THE PHYSICAL CHEMICAL PROPERTIES OF ION-EXCHANGE RESINS, *Lionel S. Goldring*, Research and Development Division, American Machine & Foundry Co., Inc., Springdale, Connecticut

ION-EXCHANGE SELECTIVITY, *D. Reichenberg†*, School of Chemistry, Rutgers, The State University, New Brunswick, New Jersey

RESIN SELECTIVITY IN DILUTE TO CONCENTRATED AQUEOUS SOLUTIONS, *R. M. Diamond*, Lawrence Radiation Laboratory, University of California, Berkeley, California, and *D. C. Whitney*, Shell Development Company, Emeryville, California, Consultant, Lawrence Radiation Laboratory, University of California, Berkeley, California

INTERPRETATION OF ION-EXCHANGE PHENOMENA, *Jacob A. Marinsky*, Department of Chemistry, State University of New York at Buffalo, Buffalo, New York

---

Current affiliations:
*Professor of Chemistry, University of Maryland, College Park, Maryland

†Ministry of Technology, National Physical Laboratory, Teddington, Middlesex, England

Volume 2

ION EXCHANGE IN GLASSES, *Robert H. Doremus*, General Electric Research and Development Center, Schenectady, New York

ION EXCHANGE IN MOLTEN SYSTEMS, *E. C. Freiling* and M. H. Rowell*, Nuclear Technology Division, U. S. Naval Radiological Defense Laboratory, San Francisco, California

THE ION-EXCHANGE PROPERTIES OF ZEOLITES, *Howard S. Sherry*, Research Department, Central Research Division Laboratory, Mobil Research and Development Corporation, Princeton, New Jersey

INTERACTIONS BETWEEN ORGANIC IONS AND ION-EXCHANGE RESINS, *Jehuda Feitelson*, Department of Physical Chemistry, The Hebrew University, Jerusalem, Israel

PARTITION CHROMATOGRAPHY OF SUGARS, SUGAR ALCOHOLS, AND SUGAR DERIVATIVES, *Olof Samuelson*, Department of Engineering Chemistry, Chalmers Tekniska Högskola, Göteborg, Sweden

SYNTHESIS OF ION-EXCHANGE RESINS, *R. M. Wheaton*[†] and *M. J. Hatch*[‡], The Dow Chemical Company, Midland, Michigan

Volume 3

EXTRACTION OF METALS BY CARBOXYLIC ACIDS, *D. S. Flett*, Mineral Science and Technology Division, Warren Spring Laboratory, Department of Trade and Industry, Stevenage, Hertfordshire, UK, and *M. J. Jaycock*, Chemistry Department, Loughborough University of Technology, Leicestershire, UK

SOLVENT EXTRACTION WITH SULFONIC ACIDS, *G. Y. Markovits* and *G. R. Choppin*, Department of Chemistry, Florida State University, Tallahassee, Florida

NUCLEAR MAGNETIC RESONANCE STUDIES OF ORGANOPHOSPHORUS EXTRACTANTS, *W. E. Stewart*, Savannah River Laboratory, E. I. duPont de Nemours & Co., Aiken, South Carolina, and *T. H. Siddall, III*, Louisiana State University, New Orleans, Louisiana

EXPERIENCE WITH THE AKURVE SOLVENT EXTRACTION EQUIPMENT, *J. Rydberg, H. Reinhardt*, and *J. O. Liljenzin*, Department of Nuclear Chemistry, Chalmers University of Technology, Fack, Göteborg, Sweden

---

Current affiliations:

*Head of the Analytical Branch, U. S. Naval Weapons Center, Code FCA, Dahlgren, Virginia

†Associate Scientist, Western Division Research, Dow Chemical Company, California

‡Professor of Chemistry, New Mexico Institute of Mining and Technology, New Mexico

Volume 4

ION EXCHANGE IN NONAQUEOUS AND MIXED SOLVENTS, *Yizhak Marcus*, Department of Inorganic and Analytical Chemistry, The Hebrew University, Jerusalem, Israel

LIGAND EXCHANGE CHROMATOGRAPHY, *Harold F. Walton*, University of Colorado, Boulder, Colorado

LIQUID ION EXCHANGE TECHNOLOGY, *Robert Kunin*, Rohm and Haas Company, Philadelphia, Pennsylvania

ELECTRONIC AND IONIC EXCHANGE PROPERTIES, CONDUCTIVITY, AND PERMSELECTIVITY OF ORGANIC SEMICONDUCTORS AND REDOX EXCHANGERS, *René Buvet*, Laboratoire d'Energetique Electrochimique, Faculté des Sciences de Paris, Paris, France, *Michel Guillou*, Laboratoire d'Electrochimie des Matériaux, Université de Rouen, Rouen, France, and *Liang-Tsê Yu*, Centre National de la Recherche Scientifique, Faculté des Sciences de Paris, Paris, France

EQUATIONS FOR THE EVALUATION OF FORMATION CONSTANTS OF COMPLEXED ION SPECIES IN CROSSLINKED AND LINEAR POLYELECTROLYTE SYSTEMS, *Jacob A. Marinsky*, Department of Chemistry, State University of New York at Buffalo, Buffalo, New York

Volume 5

NEW INORGANIC ION EXCHANGES, *A. Clearfield**, Department of Chemistry, Ohio University, Athens, Ohio, *G. H. Nancollas* and *R. H. Blessing*, Department of Chemistry, State University of New York at Buffalo, Buffalo, New York

APPLICATION OF ION EXCHANGE TO ELEMENT SEPARATION AND ANALYSIS, *F. W. E. Strelow*, National Chemical Research Laboratory, Pretoria, South Africa

PELLICULAR ION EXCHANGE RESINS IN CHROMATOGRAPHY, *Csaba Horvath*, Yale University, New Haven, Connecticut

Volume 6

ISOLATION OF DRUGS AND RELATED ORGANIC COMPOUNDS BY ION-PAIR EXTRACTION, *Göran Schill*, Department of Analytical Pharmaceutical Chemistry, University of Uppsala, Uppsala, Sweden

---

Current affiliation:
*Department of Chemistry, Texas A & M University, College Station, Texas

THE DYNAMICS OF LIQUID-LIQUID EXTRACTION PROCESSES, *G. G. Pollock*, Chevron Research Corporation, Richmond, California and *A. I. Johnson*, University of Western Ontario, London, Western Ontario, Canada

APPLICATION OF THE SOLUBILITY CONCEPT IN LIQUID-LIQUID EXTRACTION, *H. M. N. H. Irving*, Department of Inorganic and Structural Chemistry, University of Leeds, Leeds, England

SOLVENT EXTRACTION IN THE SEPARATION OF RARE EARTHS AND TRIVALENT ACTINIDES, *Boyd Weaver*, Oak Ridge National Laboratory, Oak Ridge, Tennessee

ION EXCHANGE AND SOLVENT EXTRACTION

Chapter 1

INTERPHASE MASS TRANSFER RATES OF CHEMICAL REACTIONS WITH CROSSLINKED POLYSTYRENE

Gabriella Schmuckler

and

Shimon Goldstein

Department of Chemistry
Technion-Israel Institute of Technology
Haifa, Israel

I.	INTRODUCTION	1
II.	CLASSIFICATION ACCORDING TO MECHANISM	2
	A. Shell-Progressive Mechanism	2
	B. Continuous Reaction Model	4
III.	RATE EQUATIONS FOR SHELL-PROGRESSIVE REACTIONS	4
	A. Mass Transfer Model for the Diffusion-Controlled Sulfonation Reaction	6
	a) Test of the Mathematical Model	7
	1) Film Diffusion Control	7
	2) Diffusion through the Reacted Layer	9
	B. Model of the Chemical Resistance Controlled Chloroacetylation Reaction	14
	a) Mathematical Model	17
IV.	CHLOROMETHYLATION - A HOMOGENEOUS REACTION MODEL	19
	a) Mathematical Model	21
V.	APPLICATION OF THE SHELL-PROGRESSIVE MODEL TO ION-EXCHANGE PROCESSES	25
	ACKNOWLEDGMENT	27
	REFERENCES	27

I. INTRODUCTION

The general aspects of ion-exchange resin preparations by electrophilic substitutions of cross-linked polystyrene have already been discussed in some detail [1] in this series, but relatively few studies are available on the chemical kinetics and mass transfer rates encountered in reactions on polymers.

Wheaton and Harrington [2] have stressed the importance of swelling of the copolymer for achieving uniform products of high capacity. Accordingly, swelling agents are customarily added in order to enhance the transfer rate of reactant into the rigid copolymer matrix.

For those systems in which the electrophilic substitutions are carried out on the solid copolymer the overall reaction rates are influenced not only by the rate of chemical reactions occurring in or at the surface of the solid but also by the mass transfer rates of fluids both through the solid and across the fluid film surrounding it [3].

In this chapter, kinetic models are developed for heterogeneous reactions of electrophilic reactants in the solid-fluid system. A better understanding of the rate processes involved in the interphase mass transfer phenomena encountered in these reactions will lead to better synthetic procedures, which in turn will result in products having a high capacity as well as a high degree of homogeneity.

II. CLASSIFICATION ACCORDING TO MECHANISM

The classification of heterogeneous solid-fluid reactions is dependent on the conditions of the system, such as the internal structure of the copolymer, the relative velocities of chemical reactions, and the diffusion of reactant, as well as on the geometry of the solid. In the following, two cases are considered.

A. Shell-Progressive Mechanism

In this case the porosity of the solid copolymer is small, either because of its internal structure or due to its swelling properties. The copolymer is thus practically impervious to the fluid reactant, and the initial reaction is to be visualized as taking place on the outer surface of the solid bead. The site of the reaction then moves inwards, the reaction itself occurring continuously at the surface of the unreacted core.

Slicing a partly reacted copolymer bead and examining its cross-section, the unreacted core is found to be sharply distinguished from the layer of product. Analogous heterogeneous reactions are the combustion of carbonaceous deposits [4] and the oxidation of metal spheres in oxygen [5].

For systems of this kind a set of partial differential equations with suitable initial and boundary conditions can be obtained to describe the consumption of solid reactant by the chemical reactions. Schematic representations of the model are shown in Figs. 1 and 1a.

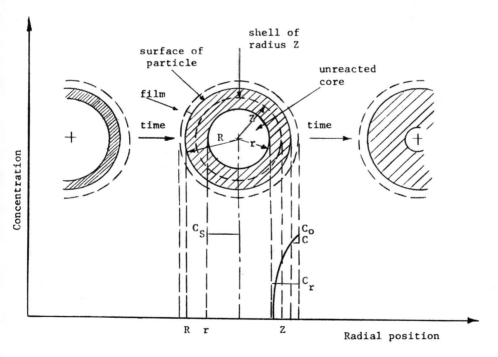

FIG. 1: Concentration profile of a partially sulfonated copolymer bead.

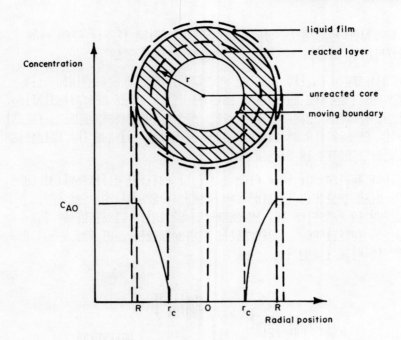

FIG. 1a: Schematic representation of the shell-progressive mechanism in a spherical ion-exchange bead.

B. Continuous Reaction Model

If the solid contains enough porosity to allow the fluid reactant to pass through freely, it is reasonable to assume that the reaction between fluid and solid occurs uniformly throughout the solid phase. Thus, solid reactant is converted continuously and progressively as shown in Fig. 2. In order to analyze the reaction kinetics in this case, partial differential equations applicable to homogeneous reactions are employed.

III. RATE EQUATIONS FOR SHELL-PROGRESSIVE REACTIONS

The shell-progressive mechanism is based on the assumption that a moving boundary divides a reacted from an unreacted zone

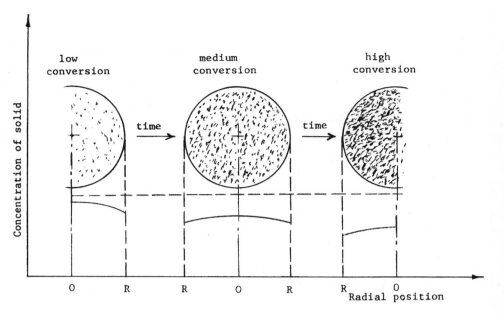

FIG. 2: Continuous reaction model (chloromethylation).

in, for example, a copolymer bead during an electrophilic reaction [6]. In this particular case, the rate at which the boundary moves inward is controlled by three consecutive processes [3]:
1) Interphase mass transfer of the reacting species from the bulk of the liquid phase to the outer surface of the solid copolymer bead;
2) Intraparticle diffusion of the reacting species from the outer surface through the reacted portion of the copolymer to the zone of reaction; and
3) Chemical reaction of the reacting species with the copolymer surface of the zone.

The relationship, "time-on-stream" vs. degree of particle substitution, can be derived by determining the rate at which an individual spherical particle reacts, while the reaction is in turn controlled, first, by the mass transfer of reactant through the bead's boundary film, then by its diffusion through the reacted layer, and finally by the chemical reaction. These different means of control are governed by the relative magnitudes of the diffusional fluxes and the rate of reaction at the aromatic sites in the copolymer.

A. Mass Transfer Model for the Diffusion-Controlled Sulfonation Reaction

The existence of a sulfonated shell region can be demonstrated by the microscopic examination of a single spherical bead of partially sulfonated crosslinked polystyrene. Such an examination was first carried out by Gilliland and McMaster [6] to yield this result. With such evidence it is reasonable to assume that a shell-progressive model should provide a good quantitative representation of the sulfonation process. Such a model can be interpreted in terms of the reactant concentration profile within the spherical bead [Fig. 1]. The electrophilic reagent, consisting in this case of 6% oleum-methylene chloride and nitromethane, diffuses into the copolymer bead and is totally consumed upon initial contact with the polymeric matrix. Thus at any given moment after the onset of the reaction, the surface of the remaining, unreacted core will react as rapidly as the reagent can diffuse to it through the layer of already substituted material. Experimental data are presented to provide an appropriate test of models proposed for derivation of the quantitative relationship between the degree of conversion, x, of the sulfonic ion exchanger thus formed, and time, with temperature (Fig. 3) and particle size (Fig. 4) as the respective variables.

In Figs. 3 and 4 accurate plots of the degree of conversion are shown for up to 400 minutes. After that period the reaction slows down appreciably - so much so that about 10 days are needed to reach complete conversion with a working temperature of 25° C, while even at a temperature of 60° C full conversion still requires 2½ days. This can be explained by the high polarity of the sulfonated product formed, which does not easily swell in the reagent mixture. Accordingly, the rate of the reagent's diffusion into the unreacted core decreases.

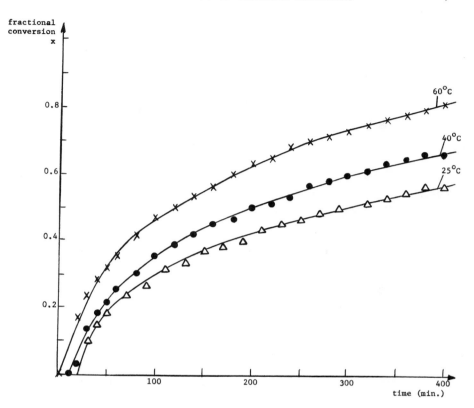

FIG 3: Effect of temperature on the kinetics of sulfonation ($\overline{R}$ = 0.0083 cm).

a) Test of the Mathematical Model

Since the chemical reaction of oleum with the aromatic sites of the copolymer is very rapid, the rate-determining step should in this case be either film diffusion or diffusion through the reacted layer. These two alternatives are examined below.

1) <u>Film Diffusion Control</u>: The analysis of the diffusion through a liquid film on the surface of a copolymer bead may conveniently depart from the assumption that the film has negligable curvature [7]. The flux (dN/dt) of the reagent's molecules through the film is given by

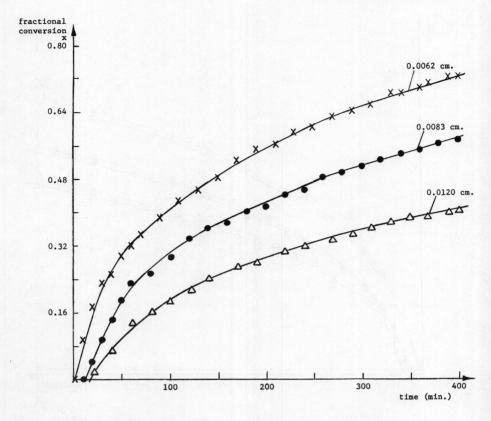

FIG 4: Effect of particle size on the kinetics of sulfonation at constant temperature (25°C).

$$\frac{dN}{dt} = 4\pi R^2 (C_o - C) \frac{D}{b} \tag{1}$$

where: D is the coefficient of diffusion through the film;
C_o is the overall concentration of the reagent;
C is the concentration of the reagent on the surface of the bead; and
b is the thickness of the film.

The concentration gradient may be taken to be constant throughout the liquid film because of the large excess of reagent. It is also reasonable to assume that the reagent molecules are consumed immediately upon reaching the bead surface. The concentration, C, is thus equal to zero and Eq. (1) re-written in the form:

$$\frac{dN}{dt} = -4\pi R^2 C_o \frac{D}{b} \qquad (2)$$

The flux, for its part, can be equated either to the decrease of volume or to the radius of the unreacted core attendant upon the disappearance of fluid reactant:

$$\frac{dN}{dt} = -C_S 4\pi r^2 \left(\frac{dr}{dt}\right) \qquad (3)$$

Equating Eqs. (2) and (3) and integrating:

$$t = \frac{C_S Rb}{3DC_o}\left[1 - \left(\frac{r}{R}\right)^3\right] \qquad (4)$$

The relation between the degree of conversion, x, and the radius of the unreacted core is given by

$$1 - x = \left(\frac{r}{R}\right)^3 \qquad (5)$$

Substituting Eq. (5) into Eq. (4) gives the relation between the degree of conversion, x, and time:

$$t = \frac{C_S Rb}{3DC_o} x \qquad (6)$$

According to Eq. (6) a plot of x vs. time should be linear, but Figs 3 and 4 show this not to be so. It can therefore be concluded that film diffusion alone is not the rate determining step.

2) Diffusion through the Reacted Layer:

The concentration gradient of reactant molecules at any shell of radius r in the reacted layer is given by (as presented in Fig. 1a)

$$\frac{dN_A}{dt} = 4\pi r^2 D_e \frac{dC_A}{dr} \qquad (7)$$

where D_e is the effective diffusivity of A through the porous reacted layer.

The initial and the boundary conditions of Eq. (7) are as follows:

at $r = R$: $C_A = C_{A0}$

while

at $r = r_c$: $C_A = 0$, - because the reactant is consumed as rapidly as it is diffused to the surface of the unreacted core.

Integration of Eq. (7) gives

$$\frac{dN_A}{dt} = 4\pi R r_c D_e \frac{C_{A0}}{R - r_c} \tag{8}$$

As the reaction progresses, the decrease of the radius of the unreacted core accompanying the consumption of liquid reactant develops as follows:

$$-\frac{dN_A}{dt} = C_{SO} 4\pi r_c^2 \frac{dr_c}{dt} \tag{9}$$

where C_{SO} is the concentration of the solid resin.

Equating Eqs. (8) and (9) and separating the variables yields:

$$-C_{SO} \int_R^{r_c} \left(\frac{1}{r_c} - \frac{1}{R}\right) r_c^2 \, dr_c = D_e C_{A0} \int_0^t dt \tag{10}$$

Integrating Eq. (10) yields the relation between t and the decrease of the radius of the unreacted core, as follows:

$$t = \frac{C_{SO} R^2}{6 D_e C_{A0}} \left[1 - 3\left(\frac{r_c}{R}\right)^2 + 2\left(\frac{r_c}{R}\right)^3\right] \tag{11}$$

In terms of fractional conversion, x [Eq. (5)], the final expression is:

$$t = \frac{C_{SO} R^2}{6 D_e C_{A0}} \left[3 - 3(1 - x)^{2/3} - 2x\right] \tag{12}$$

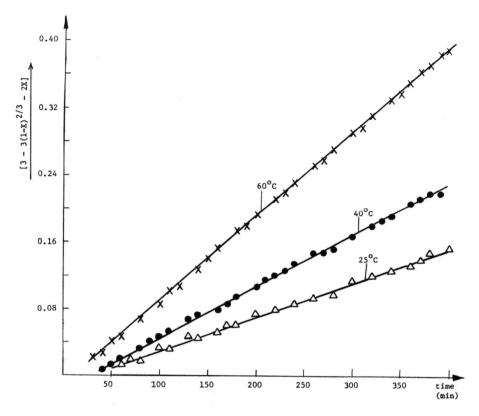

FIG 5: Test of mathematical model for sulfonation, Eq. (12).

The validity of Eq. (12) has been tested by employing the experimental values of x and t presented graphically in Figs. 3 and 4 to construct plots in Fig. 5 of $[3 - 3(1-x)^{2/3} - 2x]$ vs. t for a given bead size and three different temperatures. A similar representation of the x versus t data in Fig. 6 represents the same reaction taking place in beads of three different sizes but at one temperature. All points on the graphs were corrected by the least squares method using the CSMP computer program.

The straight lines that are obtained in all cases do not pass, as they should, through the origin because of a slight deviation from linearity near the origin. This result can be explained by the

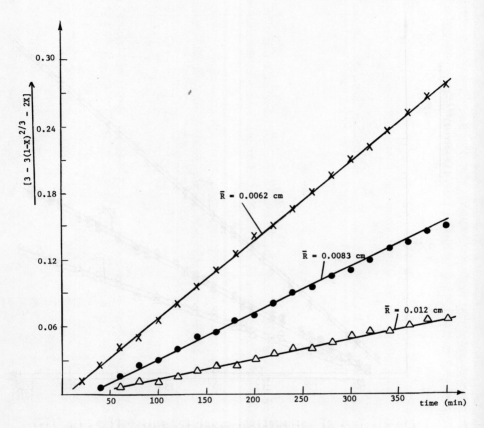

FIG 6: Test of mathematical model for sulfonation, Eq.(12).

fact that at the beginning of the reaction, i.e. after very short reaction times, the thickness of the reacted layer is still very small and thus comparable with that of the liquid film adhering to the particle. The film's resistance to the diffusion of the reactant can therefore become comparable with the resistance provided by the outer shell of the resin [8].

From the slopes of the straight lines of Figs. 5 and 6 diffusion coefficients (D_e) for the reagent through the reacted layer have been calculated as shown

INTERPHASE MASS TRANSFER RATES OF CHEMICAL REACTIONS

$$\text{Slope } k = \frac{C_{SO} \bar{R}^2}{6 D_e C_{AO}} \tag{13}$$

$$D_e = \frac{C_{SO} \bar{R}^2}{6 k C_{AO}} = \frac{cm^2}{sec} \tag{14}$$

where C_S, the concentration of aromatic rings in the polymeric matrix, is calculated from the density of the copolymer (d=1 gr/cc) and from the average molecular weight of a polymeric unit, which is 106 in the 8% crosslinked polymer:

Thus for this computation of D_e

$$C_S = \frac{1}{106} \times 10^3 = 10 \text{ mmole/cc}$$

and the concentration of the reagent, C_0, is by titration 1 mmole/cc.

The average radius of the copolymer beads, $\bar{R}$, measured with a microscope is listed together with the calculated values of the diffusion coefficient in Table 1.

Table 1: Effective Diffusivity Coefficients Through the Sulfonated Layer.

temp. °C	mesh size	mean radius ($\bar{R}$(cm.))	Diffusion coef. D_e(cm^2/sec.)
25	20-50	0.0120	7.4×10^{-10}
25	50-100	0.0083	7.7×10^{-10}
25	130-150	0.0062	7.6×10^{-10}
25	50-100	0.0083	7.7×10^{-10}
40	50-100	0.0083	11.5×10^{-10}
60	50-100	0.0083	19.2×10^{-10}

The excellent correlation in Figs. 5 and 6 of experimental data taken from Figs 3 and 4, for use in Eq. (12) to compute the ordinate of these figures is strongly supportive of the proposal that the sulfonation reaction is indeed diffusion-controlled inside the reacted layer.

Results in Table 1 indicate that the diffusion coefficient is independent of particle size but is influenced by changes in temperature. This is typical of diffusion-controlled reactions. The activation energy computed for this process from the rate constants obtained at three different temperatures is 6.1 kcal/mole, a value within the range anticipated for a diffusion-controlled reaction.

B. Model of a Chemical Resistance Controlled Chloroacetylation Reaction

The chloroacetylation of crosslinked polystyrene - which is found to proceed by yet another path, distinct from those of the more common reactions - had originally been undertaken by the authors with a view to developing a new preformed matrix for weak-base anion exchangers. The fact that

$$R - \underset{CH_2Cl}{C} = O$$

the chloromethyl group is directly attached to the polar carbonyl group should weaken the basicity of the product obtained after amination of the chloroacetylate, and thus make possible the elution of anions that cannot be eluted from anion exchangers made from chloromethylated crosslinked polystyrene. The reaction is carried out with chloroacetyl-chloride in the presence of $AlCl_3$ as Friedel-Crafts catalyst. It provides a good illustration of the kinetic behavior to be expected during the formation of a nonionic product, which swells well in non-polar solvents, such as methylene chloride. The reaction has been investigated by Kenyon and Waugh [10], who used carbon disulphide as swelling agent for polystyrene and $AlCl_3$ as catalyst; a soluble polyvinyl phenacyl chloride of low molecular weight is obtained. The mechanism of the acetylation of small aromatic molecules in the presence of $AlCl_3$ as catalyst was investigated by Brown et al. [11] and found to produce a 1:1 complex between acetyl chloride and $AlCl_3$, which is soluble in methylene chloride.

With these results available the kinetics of the chloroacetylation of crosslinked polystyrene could be analyzed. The copolymer beads, swollen in methylene chloride, were reacted for a few hours with an equimolar solution of chloracetyl chloride and $AlCl_3$ at 25°C and 40°C. The progressive conversion of the copolymer was followed; the results are presented in Figs 7 and 8 : One pair of curves represents two different temperatures with beads of a given size (Fig. 7), the other - two different sizes of beads at a

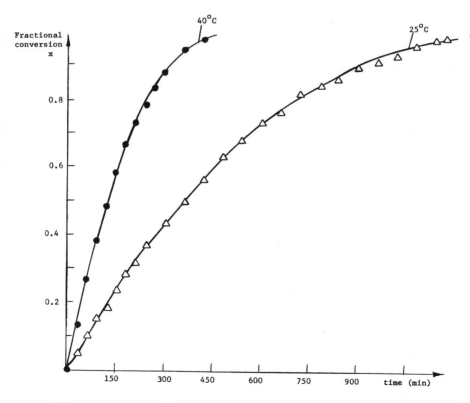

FIG 7: Effect of temperature on the kinetics of chloroacetylation (R = 0.012 cm).

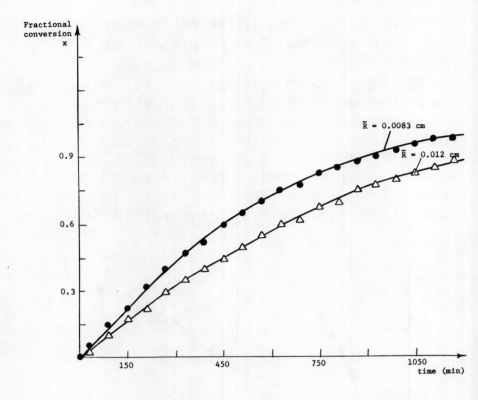

FIG 8: Effect of particle size on chloroacetylation (t = 25°C).

fixed temperature (Fig. 8). From the manner in which particle size influences the conversion - as illustrated by Fig. 8 - the "shell-progressive" model may be considered to furnish the appropriate interpretation of the experimental results. The decisive influence of the temperature on the process (Fig. 7) - the size of the beads being fixed - is a clear indication that it is not diffusion controlled. Moreover, since the chemical interaction is so much more temperature-sensitive than it is in the sulfonation process, it is reasonable to consider from the outset that the mass-transfer is controlled by surface interaction. Because of the nonionic nature of the product its swelling properties in methylene chloride

could be expected to be very good and diffusion through the reacted layer accordingly very rapid to support this estimate further.
a) Mathematical Model

With the hypothesis that the rate of reaction is controlled by interaction at the surface of the unreacted core, a rate equation can be derived that describes the data reported in Figs. 7 and 8 . Fig. 9 shows the concentration profile of the reactant over the cross-section of a spherical bead of cross-linked polystyrene; there is no concentration gradient either across the film or the product layer. The rate of reaction is proportional to the available surface of the unreacted core, and

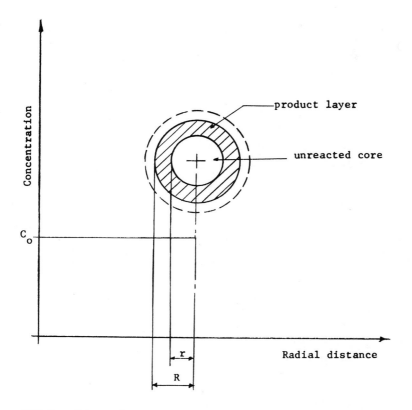

FIG 9: Schematic representation of a partially chloroacetylated copolymer bead.

thus, based on the unit surface of unreacted core - $4\pi r^2$, the rate of reaction is given by

$$\frac{dN}{dt} = 4\pi r^2 k C_S \qquad (15)$$

By equating Eq. (15) to Eq. (3) the decrease in radius of unreacted core is obtained:

$$-\frac{dr}{dt} = k \qquad (16)$$

which, on integration, yields:

$$R - r = kt \qquad (17)$$

By introducing the expression for the fractional conversion (Eq. (5)):

$$R[1 - (1-X)^{1/3}] = kt \qquad (18)$$

Plots of $1 - (1-X)^{1/3}$ vs. time for copolymer beads of average radius 0.012 cm at two different temperatures are presented in Fig. 10. The lines obtained are straight and pass through the origin, to confirm the validity of Eq. (18). The relatively great distance between them stresses the decisive influence of temperature. Such behavior is typical of chemical-resistance controlled reactions. Different bead sizes also result in straight lines (Fig. 11), a conclusive indication of the shell-progressive nature of the process.

Rate constants for chloroacetylation were determined from the slopes of the straight lines in Fig. 10. From the experiments performed by the authors at 25°C and 40°C, the activation energy is calculated to be 12.4 kcal/mole; this value is in the range characterizing a chemical reaction.

A similar analysis of experimental results was made by McKewan (12), in his investigation of the kinetics of iron ore reduction. He, too, was able to prove that the rate-determining

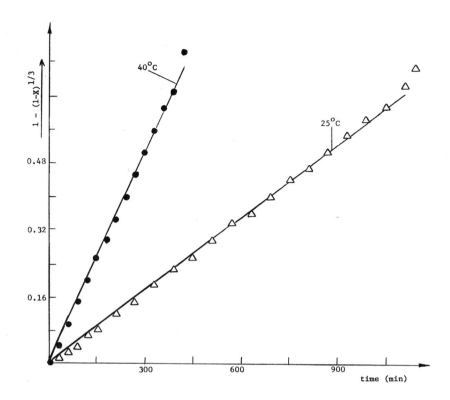

FIG. 10: Test of mathematical model for chloroacetylation (Eq. 18, $\overline{R} = 0.012$ cm).

step in that particular reaction is through chemical surface interaction.

IV. CHLOROMETHYLATION - A HOMOGENEOUS REACTION MODEL

In this catalytic Friedel-Crafts reaction, crosslinked polystyrene is reacted with chloromethyl methyl ether (CMME), which is also the swelling agent. Pepper [13] in the early fifties had already noted that the rate of chloromethylation of spherical crosslinked polystyrene beads is independent of particle size. This is a clear indication that under the good swelling conditions

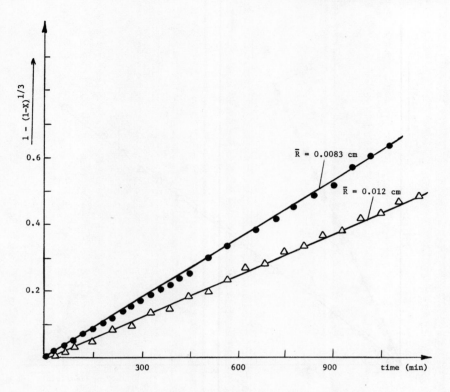

FIG. 11: Test of mathematical model for chloroacetylation (Eq. 18, T = 25° C).

existing in the reagent mixture there is no moving boundary inside the beads as described in connection with the sulfonation and chloroacetylation reactions. Rate independence of particle size is a feature of other homogeneous reactions between two different phases, e.g. the gas-solid reaction of SO_2 with calcined limestone studied by Borgwardt [20].

A homogeneous chemical model should, accordingly, apply to the chloromethylation of crosslinked polystyrene, as illustrated by Fig. 2. The kinetic equation governing this reaction should be similar to those for homogeneous systems, such as the chloromethylation of small aromatic molecules, e.g. benzene or toluene, the only difference being that the distance between the aromatic rings in the solid polymer is smaller than that in the liquid monomer and that steric hindrance must, therefore, be expected.

In order to achieve homogeneous chloromethylation of cross-linked polystyrene, the beads should be swollen in CMME, and $ZnCl_2$ should be used as a catalyst. In the first study [14] of the kinetics of this system - in which the catalytic reaction was represented by the Michaelis-Menten scheme - collection of the kinetic data was, for technical reasons, begun only after the reaction had been in progress for one hour. More recently, however, the kinetic data could be collected from the very beginning of the reaction, and are now interpreted in terms of a pseudo-unimolecular reaction.

a) Mathematical Model

The rate at which small aromatic molecules are chloromethylated is given by a second-order equation (15), according to the mechanism of the reaction [16]. Since in the chloromethylation of cross-linked polystyrene the reactant - CMME - is also the solvent, the reaction may be assumed to take place in the presence of a large excess of reactant. There is therefore only a small change in the concentration of the reactant, and the reaction is accordingly considered to be pseudo-unimolecular first order. The rate equation is

$$\frac{df}{dt} = ka(b - f) \qquad (19)$$

where f is the concentration of the product, polybenzyl chloride, at time t;
  a is the initial concentration of CMME; and
  b is the initial concentration of polystyrene units.
Integrating and converting Eq. (19) into fractional units (x=f/b) yields:

$$\ln \frac{1}{1 - x} = kat \qquad (20)$$

Eq. (20) is plotted in Fig. 12 for two temperatures, viz. 40°C and 50°C, from which it is seen that for each temperature the graph consists of two straight lines which intersect near 50% conversion. The abrupt change of slope can be attributed to the step-by-step mechanism of the chloromethylation, which is due to the steric hindrance caused by the already chloromethylated sites.

The rate equation for the case in which, at any given site on a polymer chain, the rate of reaction depends on whether the adjacent site has reacted or not has been analyzed by Alfrey [17] and Arends [18] in their theoretically based treatment of this problem.

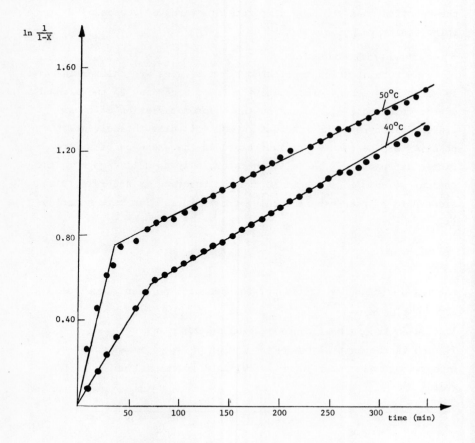

FIG 12: Kinetics of chloromethylation.

From the fact that two straight lines are obtained at each temperature studied during chloromethylation (Fig. 12) it can be deduced that two rate constants, k_1 and k_2, are sufficient to characterize the chain kinetics. Furthermore, it can be assumed that the reaction is irreversible and carried out in the presence of a large excess of reactant.

Bearing in mind the two steps of the reaction, the overall rate equation is as given by Kawabe [19], namely

$$\frac{df}{dt} = [ab/(2k_1-k_2)][2k_1(k_1-k_2)e^{-2k_1 at} + k_1 k_2 e^{-k_2 at}] \quad (21)$$

where k_1 and k_2 designate the reaction rate constants associated with a reactive site adjacent to, respectively, one unreacted and one reacted site according to the theoretical assessments of Alfrey and Arends.

Upon integrating Eq. (21) and substituting the fractional conversion, x,

$$x = 1 - \frac{k_1-k_2}{2k_1-k_2} e^{-2k_1 at} - \frac{k_1}{2k_1-k_2} e^{-k_2 at} \quad (22)$$

The experimental values of k_1 and k_2 are determined from the slopes of the lines in Fig. (12). A comparison between the calculated values of the fractional conversion according to Eq. (22) and typical experimental values is presented in Fig. (13). The good agreement between the calculated and the experimental values further justifies the application of Kawabe's rate equation to the chloromethylation of crosslinked polystyrene. A similar two-step chemical reaction mechanism is the amination of epichlorhydrin, as described by Stamberg [21]. The activation energy for that process was found to be 14.1 kcal/mole.

Activation energies and frequency factors for the reaction dis-

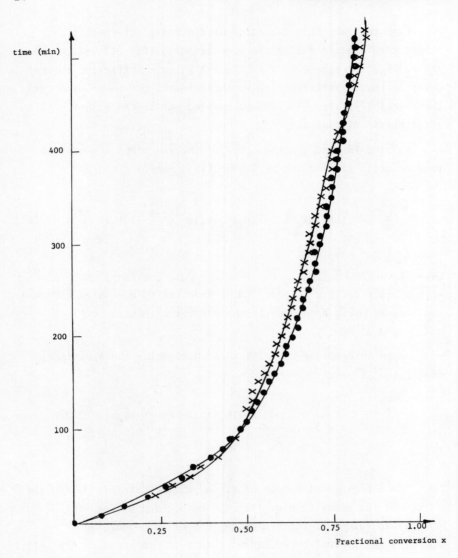

FIG. 13: Test of the two-step kinetic model for chloromethylation (Eq. 22, T = 40° C).
 x - experimental data
 ● - calculated data

cussed here, calculated from the experimentally obtained rate constants, are listed in Table 2.

INTERPHASE MASS TRANSFER RATES OF CHEMICAL REACTIONS 25

Table 2: Calculation of Activation Energies and Frequency Factors from Rate Measurements of Fig. (12).

Temp.	k(1/mole min)	E_a(Kcal/mole)	A(1/mole min)
40°C	$k_1 = 18.0 \times 10^{-4}$	23	74.0×10^{10}
50°C	$k_1 = 60.0 \times 10^{-4}$		
40°C	$k_2 = 2.2 \times 10^{-4}$	21	2.8×10^{10}
50°C	$k_2 = 6.0 \times 10^{-4}$		

The calculated values of the frequency factors for the first step are higher than those for the second step, whereas the activation energies are seen to be fairly constant. The decrease in the frequency factor is an indication of the decrease in the reactivity of the aromatic sites due to the steric effect of neighboring groups.

V. APPLICATION OF THE SHELL-PROGRESSIVE MODEL TO ION-EXCHANGE PROCESSES

The heterogeneous liquid-solid reactions discussed so far were all electrophilic substitutions of crosslinked polystyrene. There are, however, a number of ion-exchange processes which are characterized by a sharp moving boundary dividing the reacted from the unreacted parts and persisting throughout the exchange process. It seems reasonable to consider that such reactions will be governed by a shell-progressive mechanism.

The possibility of a sharp moving boundary in a spherical ion exchange bead was first suggested by Helfferich [22] for the elution of the sodium form of a weak-acid ion-exchanger with hydrochloric acid; but no experimental data were presented to confirm the hypothesis. The next investigation along these lines was made by Millar [23], who studied the kinetics of acid uptake by weak-base anion exchangers and defined three mechanisms applicable to these materials.

The kinetics of the acid uptake by weak-base ion exchangers of the type $-R_3N-$ were studied by Bolto and Kennedy [8] and were also found to follow a mechanism of diffusion through a reacted layer. Bolto and Warner [24] continued this work and extended it to the neutralization process in the weakly acidic and basic resins used in the Sirotherm process (Chapter 5), again confirming the suggested mechanism.

The most recent investigation following this line was that carried out by Selim and Seagrave [25], who studied the elution kinetics of the copper ammine complex from a sulfonic ion-exchanger.

A feature common to all the cited works on such solid-fluid interactions is the dependence of the reaction rate on the particle diameter, a characteristic of particle diffusion kinetics. In addition, the reaction rate is, of course, also dependent on the concentration of the solution.

The authors have tested a number of processes in which an ion exchanger changes from the ionized to the nonionized form. The chemical interaction is very rapid, and it is reasonable to assume rates are controlled by diffusion through the reacted layer towards the unreacted core. The following specific cases were investigated.

a). The rate of adsorption of H^+ by the sodium form of the carboxylic ion-exchanger, IRC-50.

b). The rate of adsorption of $[PdCl_4]^=$ by the chloride form of IRC-401, a strong base anion exchanger.

c). The rate of adsorption of copper ions by the sodium form of the chelating ion-exchanger, Chelex-100.

In order to test conformity with Eq. (12) (which refers to the mechanism of diffusion through a reacted layer), the degree of conversion, x, as a function of time was determined for the three cases mentioned. Graphical presentation of the data by plotting $[3 - 3(1-x)^{2/3} - 2x]$ versus time resulted, for all three cases and at three different temperatures (30°, 40°, and 50°C), in straight lines passing through the origin. These results clearly appear to confirm that the exchange rate is controlled by the diffusion through the reacted layer. Details of the experimental results will be published elsewhere later.

It is felt that the shell-progressive mechanism and the continuous reaction model represent good general approaches to the kinetics of chemical reactions on crosslinked polystyrene and in ion-exchange processes.

ACKNOWLEDGMENT

The authors are particularly indebted to Prof. E.A. Halevi for his most helpful suggestions and his lucid exposition which helped provide direction to much of the work.

REFERENCES

1. R. M. Wheaton and M. J. Hatch, Ion Exchange, Vol. 2, Ch. 6, Marcel Dekker, Inc., N.Y. (1969).
2. R. M. Wheaton and W. C. Bauman, Ind. Eng. Chem., 43, 1088 (1951).
3. O. Levenspiel, Chemical Reaction Engineering, Ch. 12, John Wiley, N.Y. (1962).
4. P. B. Weisz and R. D. Goodwin, J. Catal., 2, 397 (1963).
5. R. E. Carter, J. Chem. Phys., 34, 2010 (1961).
6. L. P. McMaster and E. R. Gilliland, Ind. Eng. Chem., Product Res. Dev., 11, 97 (1972).
7. F. Helfferich, Ion Exchange, McGraw-Hill, N.Y. (1962), p. 263.
8. R. E. Warner, A. M. Kennedy and B. A. Bolto, J. Macromolecular Sci.-Chem., 4A, 1125 (1970).
9. C. Y. Wen, Ind. Eng. Chem., 60, 34 (1968).
10. W. O. Kenyon and G. P. Waugh, J. Poly. Science, 32, 83 (1958).
11. H. C. Brown, G. Marino and L. M. Stock, J. Am. Chem. Soc., 81, 3310 (1959).
12. W. M. McKewan, Trans. Metal Soc. of AIME, 212, 791 (1958).
13. K. W. Pepper, H. M. Paisley and M. A. Young, J. Chem. Soc., 4097 (1953).

14. S. Goldstein and G. Schmuckler, Ion-Exchange and Membranes, 1, 135 (1973).
15. Y. Ogata and M. Okano, J. Am. Chem. Soc., 78, 5423 (1956).
16. G.A. Olah, "Friedel-Crafts and Related Reactions," Interscience Publishers, Ch. 21 (1964).
17. T. Alfrey and W. G. Lloyd, J. Chem. Phys., 38, 318 (1963).
18. C. B. Arends, J. Chem. Phys., 38, 322 (1963).
19. H. Kawabe and M. Yanagita, Bull. Chem. Soc. Japan, 41, 1518 (1968).
20. R. H. Borgwardt, Environ. Sci. Tech., 4, 59 (1970).
21. J. Stamberg, Coll. Czech. Chem. Commun., 29, 478 (1964).
22. F. Helfferich, J. Phys. Chem., 69, 1178 (1965).
23. G. Adams, P. M. Jones and J. R. Millar, J. Chem. Soc., A, 2543 (1969).
24. B. A. Bolto and R. E. Warner, Desalination, 8, 21 (1970).
25. M. S. Selim and R. C. Seagrave, Ind. Eng. Chem. Fundam., 12, 14 (1973).

Chapter 2

INFLUENCE OF POLYMERIC MATRIX STRUCTURE ON
PERFORMANCE OF ION-EXCHANGE RESINS

V.A. Davankov, S.V. Rogozhin, M.P. Tsyurupa

Institute of Organo-Element Compounds,
Academy of Sciences,
Moscow, USSR

I.	INTRODUCTION	29
II.	THE PREPARATION OF DIFFERENT POLYMERIC MATRIX STRUCTURES AND THEIR PROPERTIES	32
	A. Standard Styrene-Divinylbenzene Copolymer-Based Ion-Exchange Resins	32
	B. Modified Styrene-Divinylbenzene Gels and Their Ion-Exchange Resin Products	35
	1. Interpenetrating Structures	35
	2. Copolymers Obtained in the Presence of Solvating Liquids	37
	3. Telogenated Styrene Copolymers	40
	4. Styrene Copolymers Obtained in a Non-Solvating Medium	42
	C. Macronet Ion-Exchange Resins	46
	D. "Isoporous" Ion-Exchange Resins	56
	E. Macronet Isoporous Polystyrene Structures and Ion-Exchange Resins Derived From Them	59
III.	CONCLUSION	73
	REFERENCES	74

I. INTRODUCTION

In recent years ion-exchange resins have found wide application in the separation of complex organic mixtures. They are employed in isolating and purifying amino acids, antibiotics, enzymes, proteins, etc. Since the organic ions are bulky and large, and in the majority of cases, considerably exceed the size of inorganic species, the problem of enhancing the permeability of ion-exchange resins for these purposes has assumed great importance. The accessibility of the ionogenic groups for organic ions is determined largely by the structure of the resin polymer matrix.

The experimental conditions employed for the formation of polymeric gels influence their structure and, consequently, the kinetic and equilibrium properties of the ion-exchange resin products obtained. Such influence of formation conditions of the resin on the ion-exchange behavior of different organic substances is of primary importance and is the subject of this chapter.

The types of polystyrene matrices available for sorbent depend on the conditions of formation of the three-dimensional network. The final products can be divided into three major classifications. In the first styrene-divinylbenzene (DVB) copolymers are prepared by granule or block copolymerization of monomers. These matrices correspond to the standard gel. Their properties as well as those of the ion-exchange resin products based on these matrices have been thoroughly studied and it is these resins that are widely employed both in the laboratory and in industry.

The second classification is represented by styrene-divinylbenzene gels modified through monomer copolymerization in the presence of a solvent. Depending upon the role which the solvent plays in the formation of the three-dimensional network, the modified polymer matrices are further classified as:
a) polymers prepared in the presence of an inert solvent which dissolves monomers and solvates the polymer chains formed. Such a diluent decreases the number of gel chains per unit volume;
b) polymers prepared in the presence of a solvent which solvates the polymer chains and participates in the copolymerization process by playing the role of a telogen thereby shortening the mean length of the polymer chains. The polymer matrices synthesized in this manner are called telogenated polymers;

INFLUENCE OF POLYMERIC MATRIX STRUCTURE ON PERFORMANCE

c) polymers prepared in an inert solvent medium which dissolves monomers, and serves as precipitant for the growing polymer chains. The visibly porous structures formed in this manner are called macroporous structures.

Macroporous structures can also form when the divinyl-benzene-styrene polymerization takes place in the presence of an inert filler, for example, a low molecular weight polystyrene or a pulverized calcium carbonate, which is later extracted from the copolymer beads.

Another approach to a modified styrene-divinylbenzene polymer structure is through the copolymerization of monomers inside the already formed three-dimensional gel. This results in interpenetrating structures.

The third classification of matrices includes structures prepared by means of special crosslinking agents which differ significantly from DVB. They are long-chained with the distance between their two vinyl groups exceeding appreciably the full length of the DVB molecule. The ion-exchange resin with matrices so produced are called macronet polymers. They are highly permeable to large ions and the use of long-chained crosslinking agents for their preparation is extensive at this time.

Another route to ion-exchange resins with highly permeable matrices is through the introduction of additional crosslinking into the already formed DVB-styrene copolymer by means of its competitive reaction with the ionogenic group-introducing reagent. The structures so obtained are called isoporous structures.

The main objective of these matrix modifications has been to enhance the permeability of ion-exchange resins to large ions, as we have already stated. The permeability of ion-exchange resins for organic ions is monitored by the diffusion rate inside the resin granules and the maximal resin sorption capacity of these ions. The rate of diffusion provides

insight with respect to the size of channels or pores in the
swollen ion-exchange resins and determines the net rate of the
exchange reaction. The full exchange capacity of the resins for the
large ions characterizes the accessibility of these ions to the
ionogenic groups of the polymer structure. Only the combined consideration of these two factors will provide suitable estimate of
the advantage of a particular ion-exchange resin structure for the
sorbtion of organic substances.

It must be remembered when comparing the permeability of ion-exchange resins with different polymeric matrices, that the transport
of ions to the ionogenic groups and the withdrawal of their counterions is via water in the ion-exchange resin phase. Therefore one
can assess real differences in their permeability only by comparing
ion-exchange resins with similar swelling characteristics in water.

II. THE PREPARATION OF DIFFERENT POLYMERIC MATRIX STRUCTURES AND THEIR PROPERTIES

A. Standard Styrene-Divinylbenzene Copolymer-Based Ion-Exchange Resins

Standard styrene-DVB gel ion-exchange resins have been studied
in great detail. A rather complete analysis of the literature
detailing the properties of these ion-exchange resins is given in
a number of reviews (Trostyanskaya[1],Wheaton[2],Samsonov[3]). In this
chapter we consider only that information which is pertinent to the
comparison of such ion-exchange resins with the matrix-modified
ion-exchange resins.

Ordinary commercial DVB is employed in industry to synthesize
copolymers. Since this commercial product contains a great deal of
impurity which can, to some degree, participate in copolymerization,
meta- and para-divinylbenzene isomers have been used separately for
quantitative study of the copolymerization process.

A thorough study of the styrene-DVB copolymerization constants
obtained with the two isomers, [4-8], i.r. spectroscopy of the two

INFLUENCE OF POLYMERIC MATRIX STRUCTURE ON PERFORMANCE

copolymer products[9] permeation chromatography of polypropyleneglycol on the separate copolymers[10], and a careful examination of the e.p.r. spectra of the sulfonated resin product of these copolymers in the Mn^{2+} form[11] have made it possible to characterize adequately the structure of the matrix network that is formed. Styrene-DVB copolymers possess an extremely heterogeneous distribution of crosslinking bridges, which is the result of the different activities of the two monomers in the copolymerization reaction. The m- and p-DVB copolymer structures differ from one another merely in their degree of heterogeneity which is higher for copolymers with the p-isomer.

The swelling capacity of styrene copolymers with m-DVB is greater than that of polymers crosslinked with the p-isomer[12-15]. Data on the rate of sulfonation of copolymers with individual DVB isomers are extremely contradictory. Belfer[13], for example, observed that when the degree of crosslinking is higher than 7%, the more highly swollen m-DVB structures are sulfonated more easily than the p-isomer copolymer. Conversely, Wiley[16] noticed that copolymers with 8% m-DVB barely undergo sulfonation at 80°C during a four hour period whereas it is possible to introduce 4.01 mequiv/g of sulfonic acid groups into copolymers with the same amount of p-DVB. Makarova, Aptova, et al.[17] noticed no difference whatsoever in the rate of sulfonation of copolymers with 8% m- or p-DVB. Evidently, very small differences in the conditions of copolymer synthesis must strongly influence their structure and properties. The rate of sulfonation drastically falls with increase in the degree of crosslinking of the standard resins [18]. Strongly-crosslinked structures can be sulfonated only to a small degree.

The exchange capacity for organic ions (relative to Na^+ ion) of sulfonated exchangers with 2%, 5% and 10% by weight p-DVB content is presented in Table 1. All the ionogen groups are accessible for exchange only in the 2% p-DVB resin for the comparatively small tetramethyl- and tetraethylammonium ions. As the organic ion increases in size not even the full capacity of the 2% crosslinked resin is reached. Only 71% of its capacity for Na^+ ion is attained

TABLE 1

Static Ion-Exchange Capacity for Bulky Organic Ions of Sulfonated
Exchangers Based on Styrene-p-DVB Copolymer
(in % relative to Na$^+$ capacity)[19]

Organic Ion	Equilibrium Absorption of Ions p-DVB Content (wt %)		
	2%	5%	10%
$(CH_3)_4N^+$	100	86	70
$(CH_2H_5)_4N^+$	100	83	70
$(CH_3)_3C_6H_5N^+$	80	67.5	65.5
$(C_4H_9)_4N^+$	71	66	51.5

by the tetrabutylammonium ion for example. With an increase in the crosslinking the exchange capacity of these resins is further decreased.

Saldadze and co-workers[20-21] have studied the sorption kinetics of tetrabutylammonium ions on cation exchangers crosslinked with m- and p-DVB. The rate of exchange, characterized by low interdiffusion coefficients ($\log D$ = 8-10), is determined by the swelling capacity of the sulfonated resins in water. The crosslinking agent (m- or p-DVB) used in their preparation does not influence the ion exchange kinetics.

Styrene copolymers with commercial DVB swell appreciably more than copolymers with either the m- or p-DVB isomers[13,22]. Their sulfonation proceeds under milder conditions[23]. Nevertheless, the permeability of the resin products is not high; it decreases sharply with increase of the degree of polymer crosslinking. For example, the strongly-basic anion exchange resin, AB-17, containing 12% DVB by weight barely sorbs novobiocin whereas there is significant uptake of novobiocin by the less-crosslinked (2% DVB) exchanger[24]. Unfortunately, the less-crosslinked resins exhibit extremely low mechanical strength in their swollen state. As a consequence they have no practical application, in spite of their fine permeability. Moreover, they are characterized by a sharp change in volume during sorption and regeneration to complicate seriously any kind of column operation with them.

Thus, the application of the standard resins for sorbing organic substances is considerably restricted by the low permeability of resins with high and moderate degrees of crosslinking and by the mechanical and dimensional instability of the less-crosslinked resins.

B. Modified Styrene-Divinylbenzene Gels and Their Ion-Exchange Resin Products

1. Interpenetrating Structures

If a styrene copolymer containing 2% DVB by weight is swollen in styrene mixed with 2% DVB by weight, and the monomers taken up by the granules are polymerized, a product consisting of two independent networks results. Such interpenetrating structures[25] are denoted as 2×2 (according to their nominal DVB content). Each of the polymer networks contains 2% DVB so that the modified gel contains 2% of the crosslinking agent as well. However, the properties of such a polymer differ from those of the standard gel. Its density is higher, since the spaces between the primary polymeric chains are partially occupied by the second copolymer network. The swelling capacity in toluene is lower than that of ordinary styrene copolymers containing 2% DVB, coinciding with the swelling capacity of copolymers containing 4% DVB. The swelling capacity of 2×2×2 copolymers is still lower and corresponds to that of a standard gel with a DVB content of ~7%. Thus, the introduction of interpenetrating polymer networks increases their rigidity and seems to assume the function of additional crosslinking agent in the standard gel.

As is known, the content of DVB in a copolymer is often calculated from either its swelling capacity in toluene or the swelling capacity of its sulfonated product in water[26]. With the interpenetrating copolymer, however, such estimates of DVB content by these approaches are much too high. The apparent DVB content for

4×4 polymers, determined from the swelling capacity in toluene is 7.2%, whereas the content estimated from the swelling capacity of the corresponding sulfonated gel in water is 6%. For 4×4×4 copolymers these values are 14% and 7.3%, respectively. The increasing discrepancy between the two methods for estimating the apparent DVB content of these interpenetrating copolymers with increase in the degree of crosslinking is an interesting result. It is likely that during the swelling of resins in water the electrostatic forces and the considerable resultant swelling pressure introduce greater stretching of the primary three-dimensional network than the Van der Waals' interaction forces between toluene and hydrocarbon matrices.

Ion-exchange resins with interpenetrating networks possess interesting kinetic properties. Millar and co-workers[27] have compared the permeability of sulfonated exchanger with a 4.5×4.5 interpenetrating matrix and Zeo-Carb-225. The apparent DVB content in both polymers determined from their swelling capacity in water was 7%. The permeabilities of these resins with respect to Na^+ ion were indistinguishable: the diffusion coefficients in both cases were equal. With the tetraethylammonium ion the permeability of the 4.5×4.5 exchanger was higher, its diffusion coefficient corresponding to that of the standard cation exchanger with 6% DVB. On the other hand, in the 4.5×4.5 cation exchanger the high selectivity of Na^+ over H^+ ion was comparable to that of the standard highly crosslinked resins containing 15% DVB. The selectivity coefficient for the exchange of $(C_2H_5)_4N^+$ and H^+ ion in the 4.5×4.5 cation exchanger was higher than that for Zeo-Carb-225 containing 7% DVB and even higher than that for conventional ion-exchange resins containing 15% crosslinking agent[28].

We may conclude from the above that the increase in the degree of entanglement of the polymeric chains in a three-dimensional network that is endowed to the ion-exchange resin produced through interpenetration of polymers leads to kinetic properties normally associated with resins of low DVB content and to ion-exchange

INFLUENCE OF POLYMERIC MATRIX STRUCTURE ON PERFORMANCE

selectivity behavior characteristic of highly crosslinked sulfonated exchangers of standard structure.

2. Copolymers Obtained in the Presence of Solvating Liquids

Styrene-DVB copolymers obtained in the presence of toluene which effectively solvates the polystyrene chains, possess quite different properties. Millar and collaborators[29] have found that for these copolymers the swelling capacity in toluene is described by the following equation:

$$U_x = U_N + S \qquad (1)$$

where

U_x is the swelling capacity of polymer in toluene, ml/g,

U_N is the swelling capacity of ordinary copolymer with the same monomer composition in toluene, ml/g; and

S is the volume of toluene used in the reaction, ml/g.

This relation is obeyed over a considerable concentration range in the presence of the reaction medium of any quantity of crosslinking agent. From this equation it follows that the swelling capacity of copolymers in the solvating medium depends on the degree of dilution of the initial monomer mixture. The reason for such a result must be attributable to the specific conditions of formation of a three-dimensional network. When styrene-DVB copolymerization is conducted in the absence of solvent, the network which is formed, is characterized by strong entanglement of the polymer chains which limits their capacity to swell in the solvents. If the polymer chains are formed in a solvated state they are much less interlaced. When the solvent is removed from gel product containing a moderate amount of DVB, the polymer network shrinks since the flexibility of the sparsely-crosslinked segments of the polymer structure is rather high. Increase in the DVB content results in an increase in the number of densely-crosslinked regions and in the enhancement of gel rigidity. With the elimination of the solvent the rigid network does not collapse entirely[30] and part of the volume, which primarily contained the solvent, appears as a fine macropore.

The macroporous structure forms only in the presence of considerable DVB: in the range of 20-27% by weight DVB there is a sharp drop in the apparent copolymer density. These polymers are capable of absorbing not only solvating liquids like toluene, but also nitromethane, heptane, cyclohexane and water. The swelling capacity for these liquids of styrene copolymer (55% DVB) that is obtained in the presence of a two-fold volume excess of toluene[29] is:

 toluene - 2.21 ml/g
 cyclohexane - 2.10 ml/g
 heptane - 2.09 ml/g
 nitromethane - 2.01 ml/g
 water   - 1.10 ml/g

It is seen from these data that the macroporous copolymers absorb organic liquids in considerably greater quantity than water. The absorption of water is not accompanied by an increase in the polymer volume and is due only to the filling of macropores. The water value of 1.10 ml/g thus characterizes the free pore volume in the polymer. The amount of absorbed heptane or cyclohexane exceeds considerably the free volume of pores, and approaches the swelling capacity of the gel in toluene. Obviously, these solvents can penetrate not only into the polymer pores, but are also capable of solvating the polystyrene chains to induce swelling of the structure.

The inner specific surface area of the macroporous copolymers has been determined by measuring the heat of solvation with benzene. For polymers containing from 30 to 60% DVB the products obtained in two volumes of toluene, are characterized by an inner specific surface area varying from 50 to 200 m^2/g. The mean diameter of the pores has been determined by electron microscopy to be 350 Å.

In comparison with standard styrene-divinylbenzene resins polymers obtained in a solvating medium have a structure which is

INFLUENCE OF POLYMERIC MATRIX STRUCTURE ON PERFORMANCE

more accessible to chemical reagents. Toluene-modified polymers with 16% DVB sulfonate much more easily than ordinary polymers with 8% DVB[31]. Study of the permeability of the sulfonated exchangers for organic ions has shown that they possess better kinetic properties than the ordinary resins. Millar et al.[32] synthesized sulfonated exchangers containing 7, 15 and 27% DVB; they were endowed with similar swelling capacities in water, by performing their polymerization in the presence of an increasing amount of toluene. The data presented in Table 2 for these exchangers and two others with higher DVB content show that the rate of diffusion of $(C_2H_5)_4N^+$ is approximately the same in these ion-exchange resins. The diffusion coefficient of $(C_4H_9)_4N^+$ in the resins containing 34 and 55% DVB (visibly porous) is, however, one order of magnitude larger than in the exchangers containing 7 and 15% DVB. Such enhancement of the sorption rate of organic ions that these porous structures provide is a desirable aspect of these materials. However, this advantage is counterbalanced by a loss in capacity for these ions. For example the porous resins containing 55% crosslinking agent have a capacity related to the Na^+ ion of 90% for the $(CH_3)_4N^+$ and only 63% for the $(C_4H_9)_4N^+$ ion whereas the capacity of the conventional ion exchange resins containing 7% DVB for these ions is 100 and 92%, respectively. In other words, only those ionogenic groups which are located on or near the macroporous surface are accessible for large ions in the macroporous structures.

A very interesting feature of these macroporous exchangers is that its volume in water is the same as the volume of the initial polymeric matrix in toluene. For ordinary resins no such equality is observed[29]. The fact that the swelling capacity of polymers does not depend on the kind of force to which it is exposed, either electrostatic or Van der Waal, is supportive of the concept that no entanglement of polymer chains, which occurs in ordinary resins, is present in the internuclear chains of the macroporous material. (In this case one should not forget that the high polymer rigidity is caused by the high content of DVB).

In conclusion it should be noted that the sulfonated exchanger, the matrices of which are obtained in the presence of toluene, possess, along with a remarkable permeability for organic ions, a high osmotic stability. In addition, their swelling capacity at sufficiently high crosslinking does not noticeably depend on the counterion form in the case of H^+, Na^+, K^+ or Li^+.

3. Telogenated Styrene Copolymers

In the process of styrene-divinylbenzene copolymerization in a medium of toluene the latter, to some extent, can perform the role of a chain transmitter. But its contribution as a telogen to the structural aspects of the networks is minor (the chain transmission constant for homopolymerization of styrene in a toluene medium is as low as 1.25×10^{-5} [33] and there is no doubt that the main role of toluene is the solvation of the polymer chains and the dilution of the reaction medium. If copolymerization is carried out in the presence of an active telogen the forming polymer will contain shorter chains and the three-dimensional structure will be, on the whole, more friable [34].

Telogens evaluated on the basis of their chain transmission capability in the homopolymerization of styrene are listed below in the order of their relative activity:

CCl_4 > diethylbenzene > ethylbenzene > toluene

Trostyanskaya et al. [35] have compared the swelling capacity of copolymers of styrene containing 3% DVB, synthesized in the presence of these telogens. Using 15% of the above-mentioned solvents the swelling capacity of the copolymer products correlates with telogen activity.

Of all the telogens employed the most active is carbon tetrachloride. The higher the content of CCl_4 in the initial monomer, the higher is the swelling capacity of the final polymer. The content of chlorine in the copolymer increases simultaneously and reaches 2 to 5% [36,37] to confirm the participation of CCl_4 in the transmission of the kinetic chain.

INFLUENCE OF POLYMERIC MATRIX STRUCTURE ON PERFORMANCE

The permeability of the telogenated styrene copolymers has been determined by measuring the maximum molecular weight of polystyrene capable of diffusing inside the swollen gel[34]. Copolymer of styrene containing 2% p-DVB, obtained in the presence of 60% CCl_4, has a structure accessible to polystyrene with a molecular weight as large as 68000. The same copolymer obtained in the absence of a solvent is permeable only to a polymer with a molecular weight of 32000 or less.

Polymers synthesized in the presence of telogens can be sulfonated without preswelling in dichlorethane. An 85% sulfonation is reached for polymers containing 3% DVB when in CCl_4 medium; with diethylbenzene and ethylbenzene and with toluene 60% and 50% sulfonation is reached. Standard styrene copolymers with the same amount of crosslinking agent sulfonate only up to 18-22%.

Sulfonated exchangers derived from telogenated copolymers swell in water more strongly than the standard resins[35]. Their sorption capacity for an organic ion with a molecular weight of 450 depends on the amount of telogen used in the copolymerization. It rises sharply as the content of CCl_4 increases from 0 to 10%, and much more slowly on further increase in the quantity of CCl_4 to 60% [36]. It is apparent from graphical representation of such data[36] that the sorption capacity of the telogenated cation exchange resins for organic ions is higher than that of ordinary resins of the KY-2 type in an equivalent swollen state.

Trostyanskaya et al.[19] have compared the permeability of the trimethylcetylammonium ion in telogenated and standard resins. They have established that at equivalent DVB content the telogenated structure is more accessible to such a large ion. Nevertheless, the permeability of the telogenated resin is still insufficient for practical use. The maximum degree of saturation of resins with $(CH_3)_3C_{16}H_{33}N^+$ after 240 hours is 20% for the standard resin containing 2% p-DVB, whereas for the telogenated resins obtained in the presence of 40% CCl_4 it is 32%. An increase in the DVB content of the initial copolymer to 10% decreases the capacity of standard resins by as much as 15%; the telogenated resins by as much as 20%. It should be noted, however, that the choice of trimethylcetylammonium ion for studying the permeability of resins has proved to be a rather poor one. This ion is a surface-active agent and can form a lipid like shell[38] thereby preventing deep penetration of other ions into the resin granule. Conversely, studying the kinetics of exchange of Na^+ ion with tetracyclin has shown that the rate of diffusion of this antibiotic in a telogenated cation exchanger obtained from a styrene - 4.5% DVB copolymer in 50% CCl_4 is more than one order of magnitude higher than in a Dowex 50×2 cation exchanger with the same water content[39].

A great many papers have been devoted to study of the influence of another telogen, namely, diethylbenzene on the styrene-DVB copolymer structure, since this compound is one of the major impurities in commercial DVB. With an increase in the amount of diethylbenzene added to the polymerizing mixture the swelling capacity of the copolymer increases[34] and a more friable network is formed. This is inferred from the fact that the rate of introduction of ionogenic groups into such structures increases and the permeability for tetrabutylammonium ions of the sulfonated exchangers derived from these structures rises[40]. However, the influence of diethylbenzene on the copolymer properties is somewhat less than that of CCl_4. It should be noted, however, that the introduction of a small amount of diethylbenzene into the initial styrene-DVB mixture enhances the mechanical strength of the sulfonated exchangers[41,42].

4. Styrene Copolymers Obtained in a Non-Solvating Medium

When a medium which does not solvate the growing polymer chains is present in the monomer reaction mixture highly irregular microheterogeneous three-dimensional structures are formed. In a dry state they possess real porosity, even when the crosslinking agent is present in small quantity. The special features in the formation of macroporous styrene-DVB copolymer structures in such a medium are discussed below.

In the presence of non-solvating medium such as heptane, isooctane alcohols, etc. the formation of macroporous structure proceeds in several stages[43]. In the initial stage of copolymerization the reaction mass is a solution in an inert diluent of monomer, growing linear and branched oligomers, and an initiator. As the polymer chains grow and the amount of free monomer diminishes, the solvating potential of the medium gradually falls. Its content in the growing polymer cluster decreases, polymer conglomerates appear, and the system becomes microheterogeneous. The oligomers occupy the intermediate position between two phases. After accumulating on the surface of the growing crosslinked or branched polymer they stabilize the system. By the time of gel formation practically none of the medium is in the growing crosslinked polymer phase. Polymerization of the remaining monomer leads to the formation of bonds between these crosslinked particles. As a result the medium concentrates locally between the polymer microspheres, and this elimination of the solvent leads to the formation of a macroporous structure. The microspheric structure of macroporous polymer agglomerates first suggested by Kun and Kunin[44] has been confirmed by diffraction patterns obtained with the electron microscope for the Amberlite IRA-938 cation exchanger[45].

This special feature of solvent rejection during polymer structure formation defines the properties of the macroporous gels and the resins obtained from them. When one compares the swelling capacity of these gels with the swelling capacity estimated with Eq. (1), previously shown to be applicable for toluene-modified copolymers, it is quite apparent that the macroporous gels swell more extensively than predicted. Only when a small quantity of solvent is used does this equation describe the swelling behavior[46].

Macroporous copolymers have been examined very carefully. The influence of the quantity of crosslinking agent and the quantity and nature of the non-solvating medium on the porous polymer structure (the inner specific surface area, the pore volumes, the distribution of the pores along the radius, etc.) has been thoroughly studied by Jergozhin[47], Seidl et al. [48,49].

Macroporous styrene copolymers have found wide application as a stationary phase for gas[50,51] and gel[52] chromatography. Macroporous ion-exchange resins derived from them possess high exchange capacities and excellent osmotic stability[53]. Thanks to their large porosity and considerable inner surface area, they can sorb organic substances from non-polar media at a high rate[54,55].

By varying the amount of diluent and crosslinking agent, one can obtain macroporous resins with various pore diameters thereby affecting their permeability. Brutskus et al.[56] have studied the kinetics of exchange of hydrogen ions relative to $(CH_3)_4N^+$, $(C_2H_5)_4N^+$, $(C_4H_9)_4N^+$ in sulfonated macroporous exchangers. At a constant degree of crosslinking (20% DVB) the rate of diffusion of the quarternary ammonium ions increases with the increase in the amount of n-heptane introduced into the initial monomer mixture. Conversely, at a constant concentration of n-heptane increase in the degree of crosslinking from 40 to 60% leads to a decrease in the rate of sorption of the tetrabutylammonium ion. However, the rate of diffusion of the alkylammonium ions in macroporous resins are always much higher than in standard resins with the same content of DVB. The sulfonated macroporous exchangers, unlike the standard resins, do not reveal a sharp decrease in the rate of absorption with an increase in the diameter of the sorbed ion.

Table 3 presents a comparison of the kinetic characteristics of standard and macroporous cation exchangers relative to methylene blue ion[57]. The macroporous cation exchangers absorb methylene blue faster and more completely than standard gel resins with the same content of crosslinking agent. It follows from Table 3 that with the same concentration of monomer mixture the increasing degree of crosslinking in the polymer framework exerts little influence on the rate of diffusion of the dye in the macroporous resin, though its capacity for the dye decreases sharply. A macroporous cation exchanger with an inner surface area of 148 m^2/g exchanges 54% of

TABLE 2

Permeability of Toluene-Modified Ion-Exchange Resins for Organic Ions[32]

Amount of DVB %	C_s	$(CH_3)_4N^+$ Q	$(CH_3)_4N^+$ $D\times10^6$	$(C_2H_5)_4N^+$ Q	$(C_2H_5)_4N^+$ $D\times10^6$	$(C_4H_9)_4N^+$ Q	$(C_4H_9)_4N^+$ $D\times10^6$
7	0.20	100	1.08	100	0.26	92	0.015
15	0.20	-	-	100	0.29	82	0.032
27	0.22	-	-	87	0.23	67	0.088
34	0.40	95	0.49	90	0.32	65	0.24
55	0.54	90	0.62	80	0.42	63	0.32

C_s - Swelling capacity of cation exchangers in water, g/meq;
Q - Maximum capacity of organic ion in % relative to Na^+;
D - Interdiffusion coefficient of H^+-organic ions, cm^2/sec.

TABLE 3

Permeability of Macroporous and Standard Sulfonated Exchangers to Methylene Blue[57]

Amount of DVB %	macroporous cation exchangers S	macroporous cation exchangers Q	macroporous cation exchangers D	standard cation exchangers Q	standard cation exchangers D
2	-	-	-	67	1.8×10^{-9}
4	7	73	5.8×10^{-9}	60	2.2×10^{-9}
8	12	70	4.2×10^{-9}	48	3.4×10^{-10}
15	-	-	-	31	1.0×10^{-10}
20	65	70	2.8×10^{-9}	30	4.7×10^{-10}
30	100	55	2.3×10^{-9}	12	5.7×10^{-12}
40	148	54	5.5×10^{-9}	-	-
60x	223	49	6.2×10^{-11}	-	-

Remarks: Macroporous resins are produced in the presence of 100 volume % of n-heptane, with the exception of x -dilution of 60 volume %.
Q - Maximum degree of saturation of methylene blue (in%) relative to Na^+;
D - Interdiffusion coefficient, cm^2/sec;
S - Inner specific surface area, m^2/g.

its Na^+-ion capacity for methylene blue. This result shows that the exchange of large ions in macroporous structures proceeds easily only on the macropore surfaces [58]. As was stated above Millar has come to the same conclusion by studying the kinetic properties of macroporous cation exchangers whose polymer framework was obtained in a toluene medium.

C. Macronet Ion-Exchange Resins

Macronet ion-exchange resins are obtained by copolymerization of monovinyl compounds with diolefins whose linear dimensions exceed those of the divinylbenzene molecule. On swelling these ion-exchange resins, polymer chains are separated a considerable distance by the long crosslinking bridges. Thus diffusion of organic ions in the inner channels is much easier.

Synthesis of such ion-exchange resins was primarily carried out with polymers derived from the aliphatics. Only recently have macronet resins of the polystyrene type been obtained.

In the first series of studies on the synthesis of macronet resins, N,N'-alkylenedimethacrylamides were used as the long-chained crosslinking agents, according to the formula:

$$CH_2 = \underset{CH_3}{C} - CONH(CH_2)_n NHCO - \underset{CH_3}{C} = CH_2 \text{ where } n = 2, 6, 10 \text{ [59-62]}$$

Monoolefins employed were methacrylic acid[63] or its derivatives carrying different ionogenic groups such as sulfonic acid[64], phosphoric acid[65] or amino groups with different basicities [66]. The copolymerization of the monomers was carried out in acetic acid solutions of different concentrations[67] or in organic solvents like methanol and dimethylformamide[68].

The physico-chemical properties of macronet resins depend on the conditions under which the three-dimensional polymer forms. Samsonov et al.[67] obtained ion exchangers when copolymerization of methacrylic acid with 0.5 to 10 mol % of N,N'-alkylenedimethacrylamide was carried out in 95% acetic acid. These resins swell in water to a greater degree than the resins obtained in 30 to 60% aqueous solutions of CH_3COOH. The bulk density of these resins is 2 to 2.5 times lower than that of resins obtained in diluted acid; their inner specific surface area determined by low-temperature nitrogen sorption is 0.1 m^2/g. Study of the permeability of the Na^+-form of carboxylic acid exchangers crosslinked with N,N'-hexa-methylenedimethacrylamide (HMDMA) for streptomycin ion has shown[69] that with a content of HMDMA up to 10 mol % the diffusion coefficients in resins obtained in 96% acetic acid are 2 to 3 times lower than in ion-exchange resins synthesized in diluted acids. This difference disappears only when the content of the bridge-forming component reaches 10%. It is likely that the degree of solvation of the polymer chains during the formation of the three-dimensional network considerably influences the properties of the ion-exchange resin product.

The swelling capacity of macronet ion-exchange resins in water depends on the size of the crosslinking agent molecule and the degree of dilution of the initial monomer mixture. With a constant monomer concentration in solution, the swelling capacity of copolymers of the Na-salt of 4-sulfophenylmethacrylamide (SPMA) increases as the ethylenedimethacrylamide (EDMA) is replaced by hexa- (HMDMA) and deka-methylenedimethacrylamide (DMDMA)[64]. On the other hand, the swelling capacity of SPMA containing 17% of HMDMA in water is higher when the initial monomer mixture is diluted.

Macronet ion-exchange resins swell in water to a considerably greater degree than standard resins (KY-2 and KB-4) containing the same amount of DVB[67-70]. Since the water vapor sorption isotherms of macronet resins are S-shaped, Samsonov et al.[63,71-74] have attempted to calculate the distribution of pores along the radii in the macronet ion-exchange resins, using a capillary-based model. The sorption of water, however, is accompanied by swelling of the ion-exchange resin, and, consequently, by changes in the sizes of the inner channels. A more appropriate approach to the study of the inner pore sizes of swollen ion-exchange resins is believed to be provided by investigation of the resin permeability for ions of different dimensions. Such research[75] has been carried out. The permeability of sulfonic macronet resins containing various amounts of ethylene- and hexamethylenedimethacrylamides for NH_4^+, $(CH_3)_4N^+$, $(C_2H_5)_4N^+$, $(C_4H_9)_4N^+$, $(C_2H_5)_3C_6H_5CH_2N^+$, methylene blue and crystal violet was measured. Estimates of pore size distribution were made from the dependence of the relative exchange capacity of the resin on the sorbed ion diameter. It appeared that pronounced maxima in the distribution of pore diameters ranging from 2.5 to 7.0 and 10 to 15 Å are typical of macronet ion-exchange resins containing 23% EDMA. The absence of pores with diameters less than 10 Å was noticed in the structure of sulfonic macronet resins containing 9% HMDMA or EDMA. It is likely that a major fraction of the ionogenic groups in these less-crosslinked resins is located in channels whose diameter exceeds 15 Å.

A great amount of research[39,59,69,76-80] has been devoted to study of the permeability of macronet resins for alkylammonium ions and different antibiotics. The data obtained to show the influence of the length of the crosslinking agent on the permeability of the macronet ion-exchange resins are almost free from contradictions. The rate of sorption of streptomycin and methylene blue increases on substituting HMDMA for EMDA, but drops when dekamethylenedimethacrylamides are used as the crosslinking agent[69]. This phenomenon has been attributed[81] to either the collapse of the longest crosslinking agent chains or to the formation of nodules at intersections of the polymer and crosslinking agent strands which are equivalent to an additional crosslink. Such an explanation of these results is inadequate since both factors should have a similar effect on the ion-exchange resin swelling capacity. However, the swelling capacity of the macronet resins increases monotonously (uniformly) with increase in the length of the crosslinking agent. In addition, no anomolous behavior is observed in the sorption of tetracyclin. The rate of diffusion of this antibiotic increases in the macronet resins as the length of their crosslinking agent increases[82].

INFLUENCE OF POLYMERIC MATRIX STRUCTURE ON PERFORMANCE 49

With increase in the degree of crosslinking, the rate of diffusion of antibiotics falls[76], whereas the energy for activating the process increases[78]. These changes are less pronounced than for the carboxylic acid exchange resin, KB-4, or for Dowex 50 which contain a shorter crosslinking agent such as divinylbenzene.

Comparison of the kinetic characteristics of sulfonic exchange resins with different chemical properties and different matrix structures, but with the same swelling capacity in water has shown that the diffusion coefficients of antibiotics (the tetracyclin series) are one order of magnitude lower in resins crosslinked with DVB than in the macronet ion-exchange resins [83].

Musabekow et al.[84] have compared at similar degrees of swelling in water the permeability of tetrabutylammonium and oleandomycin in Dowex 50 with their permeability in resins prepared by copolymerization of the Na-salt of 4-sulfophenylmethacrylamide in combination with HMDMA (Fig.1). The macronet cation exchangers are satisfied completely by the tetrabutylammonium ion even when the degree of swelling in water is low. Conversely, the weakly-swollen Dowex 50 resins are not saturated by this ion which only replaces 60% of their Na^+-ion content. The difference in the permeability of these resins for this ion is removed only when the swelling capacity exceeds 10. The great difference in resin structure is revealed even more effectively by the sorption behavior of oleandomycin; over the complete range of resin swelling capacity the sulfonic macronet cation exchangers are much more accessible than Dowex 50 to antibiotics.

These large differences in resin permeability when comparisons are made on the basis of swelling capacity, must be analyzed further to establish the true implications of such results. For example, the macronet resins swell much more than the sulfonated polystyrene resins even though their exchange capacities are considerably lower than that of Dowex 50. On increasing the content of HMDMA from 5 to 41% in the macronet resin, its capacity drops from 3.58 to 2.27 meq/g, while for the standard sulfonated polystyrene resins containing up to 10% DVB the static exchange capacity varies from 4.5 to 5.2 meq/g. Even at a capacity as low as 2.27 meq/g the sulfonic macronet

TABLE 4

Swelling Capacity of Sulfonated Polystyrene Exchangers in NH_4^+-Form[85]

Amount of DVB (wt%)	SEC meq/g	Volume of moist resin in ml per g of dry resin	Volume of moist resin in ml per meq of exchange groups
0.5	2.75	2.9	1.05
0.5	4.75	5.56	1.27
0.5	5.3	7.5	1.40
10	2.95	1.25	0.43
10	4.75	1.5	0.32
10	5.15	1.5	0.29

resin absorbs 3 to 5 times more water per equivalent than standard resins containing up to 8% DVB[73]. This considerable difference in the swelling capacity of the cation exchangers is attributed to the difference in the linear dimensions of the cross-linking agent and to the fact that the macronet resin structure forms in the presence of a solvate which encourages the formation of a less intricate network. Undoubtedly, another sizeable contributing factor is the smaller concentration of ionogenic groups; solvation of ions at low concentration is more extensive. From the data in Table 4 one can see that there is a complicated relationship between the exchange capacity of the cation exchangers and their swollen volume. With a decrease in the capacity of a resin of 10% DVB content by almost a factor of two there is a compensating increase in the amount of water per meq. of ionogenic groups. The net result is that the swollen volume of the resin is almost the same. Hence, measurement of the swelling capacity value alone does not always provide sufficient information for a valid estimate of the situation. For a more informative analysis of the permeability of ion-exchange resins having different structures and different capacities than is presented above the permeability of resins at

INFLUENCE OF POLYMERIC MATRIX STRUCTURE ON PERFORMANCE

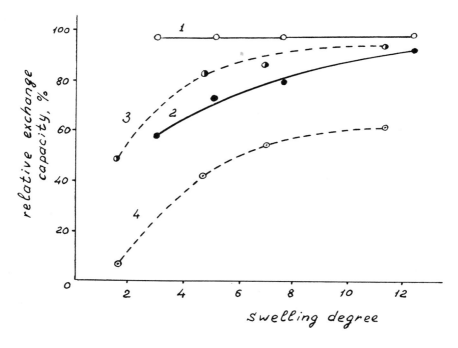

Fig. 1. Plot of exchange capacity of macronet and Dowex 50 resins for tetrabutylammonium ion (curves 1 and 3) and oleandomycin (curves 2 and 4) relative to Na^+ ion (in %) against their swelling capacity.
1,2 - macronet resins,
3,4 - Dowex 50

the same swelling capacity per equivalent of exchange groups must be compared.

When the data in Fig. 1 are reexamined by expressing swelling capacity in ml/meq, the exchange capacity curves describing the permeability of tetrabutylammonium in the macronet and standard resins converge. The difference in the resin accessibility for oleandomycin is also significantly smaller. Earlier differences observed in kinetic studies are also minimized when water content is based on the exchange capacities of the resins for interpretation of such data.

In evaluating the sorption efficiency of organic molecules the rate of diffusion of the molecule as well as the selectivity of the sorption process are considered. It is often the case that improvement of the kinetic properties of the ion-exchange resin leads to a reduction in its exchange selectivity. However, the macronet ion-exchange resins provide both adequate permeability and a high specificity for some organic ions, for example, dyes or antibiotics [77,79,80].

Kolomyetsev et al.[86-88] have studied the nature of crosslinking-bridge distribution in HMDMA copolymers with some amino derivatives of methacrylic acid. By examining the copolymer composition it was possible to establish that only one double bond of the bridging component participates in the reaction at an early stage of copolymerization. This made it possible to consider the copolymerizing system as a binary mixture. The relative reactivity ratios (r') of the comonomer vinyl groups show that only iodine N,N,N-trimethyl-aminoethylmethacrylamide (ITMAE) and methylsulfate of N,N,N-trimethyl-amino-p-phenylmethacrylamide (MSPA) with HMDA ($r_1' = 1.76$, $r_2' = 0.56$ and $r_1' = 2.08$, $r_2' = 0.56$, respectively) are characterized by an equal distribution of the crosslinking bridges since the copolymerization constants of these monomers satisfy rather well the azeotropic condition $0.5\ r_1' = 2$, $r_2' = 1$ [89].

In the copolymerization of N,N-dimethylaminoethylmethacrylamide (DMA), N,N-dimethylamino-p-phenylmethacrylamide 2-methyl-5-vinyl-pyridine and methylsulfate of N-methyl-2-methyl-5-vinylpyridine (MSVP) with HMDMA, the polymer at early stages of the process is either depleted of crosslinking agent or the crosslinking agent is involved in the general ion-exchange resin structure in the form of strongly-crosslinked entities which do not carry the ionogenic groups.

The difference in distribution of the crosslinking bridges affects the permeability of the macronet anion exchangers. The ion-exchange resins obtained from HMDMA copolymers with ITMAE and MSPA possess the highest relative capacity for dyes with ionic diameters of 23 to 30 Å although their swelling capacity in water is lower than the swelling capacity of the anion exchangers obtained from MSVP and DMA at the same degree of crosslinking[88].

An important drawback of macronet ion-exchange resins is the low chemical stability of their polymer matrix; the amide bridging components undergo hydrolysis in acid media[66]. In addition, all the ion-exchange resins described above are produced as irregular particles.

In recent years a variety of papers concerned with the production of granular macronet ion-exchange resins obtained from styrene copolymers has appeared in the literature. These papers describe only macronet copolymer synthesis techniques and do not present any data on the properties of the ion-exchange resins themselves. These procedures are described in detail by Ergozhin and Kurmanaliyev[90]. Our discussion of these materials is therefore confined to a brief summary of the synthetic routes to macronet polystyrene exchangers.

Table 5 presents a list of the more important types of crosslinking agent used in the synthesis of macronet styrene copolymers. Copolymerization is carried out both in the absence of solvent (for example, in the case of N,N'-alkylenedimethacrylamides [91] and in media which solvate the polymer chains in different ways (for example, copolymerization with aromatic bis-phenyl esters [92]. The swelling capacity of copolymers in organic solvents is determined by the type of crosslinking agent, its quantity and the concentration of the monomer mixture.

A detailed study of the chloromethylation of macronet styrene copolymers has shown that it is easy to introduce 23 to 25% chlorine into such polymers[92,93]. By aminating the chloromethylated copolymers with amines of different basicities, a series of anion exchangers has been produced[94-98]. Ergozhin et al.[97] has noted that the application of long-chained crosslinking agents of the alkylenedimethacrylamide type leads to the formation of a friable polymer

TABLE 5

Macronet Styrene Copolymers

General formula: for divinyl monomer	R_1		References
$CH_2 = C - R$			102,103
CO			
O			92,94
R_1			92,94
O			
CO			93,106
$CH_2 = C - R$			93
R - H; CH_3			93
			93,100,106
$CH_2 = C - R$			93,98,106,107,109
CO			
NH			95,98,106,109
R_1			
NH			96,98,106,107,109
CO			
$CH_2 = C - R$			96,98,106,107,109
			107
R = H; CH_3	$(CH_2)_n^-$	n=2,6,10	91,95,97,106,108
$CH_2 = CH$		n=1,6,10	18
R			18,170
$CH_2 = CH$			164

matrix structure, which allows amination without pre-swelling of the chloromethylated copolymers. The introduction of ionogenic groups into the macronet copolymers is carried out faster and more completely than with the styrene-DVB copolymers in the gel and macroporous conformations[92,93].

Copolymers with dimethacrylic esters of 2,2-bis(p-oxyphenyl)-propane, 1,12-(p-oxyphenyl)dodecane and 1,4-dioxybenzene have been synthesized[99-101]. The diffusion coefficient of sudan III in the swollen copolymers containing a small quantity of crosslinking agent depends little on the crosslinking agent length. The coefficient is barely different from the diffusion coefficient of dyes in copolymers of styrene and technical grade DVB at the same swelling capacity as polymers in toluene (2-3vol/vol). However, at a smaller swelling capacity, copolymers containing 12 methylene groups between the benzene rings of the crosslinking bridge possess a permeability one order of magnitude greater than that of 1,4-dioxybenzene dimethacrylic ester copolymers[99].

Kurmanaliyev et al.[92] as well as Trushin et al.[102-105] have attempted to combine the properties of the macronet and macroporous structures by carrying out the copolymerization reaction in a non-solvating medium. However, these authors say nothing about the advantages that such a resin may have over true macronet or macroporous ion-exchange resins.

Macronet styrene copolymers contain labile amide or ester groups in the crosslinking bridge. They therefore cannot react with the ionogenic-introducing reagent under rigid conditions. Thus, sulfonation by sulfuric acid at 100°C of styrene copolymers with bis-acrylamide derivatives of benzene, diphenylmethane, diphenyloxide and diphenylsulfide[106] and HMDMA[69] is accompanied by partial hydrolysis of amide bonds. When the crosslinked copolymer is subjected to pre-swelling in nitrobenzene, the sulfonation process leads to the production of a completely soluble product. The authors were able to obtain a monofunctional cation exchanger with a capacity of 4.8-5 meq/g by sulfonating the polymer, swollen in dichloroethane, with oleum at 25°C. Nevertheless, there is no doubt that such cation exchangers are of inadequate stability. This considerably lowers the value of macronet polystyrene ion-exchange resins.

D. "Isoporous" Ion-Exchange Resins

It follows from the data considered earlier that it is possible to improve appreciably the permeability of ion-exchange resins by means of different styrene copolymer modifications. Yet not one of the above-described methods for altering the polymer matrix structure changes the nature of the crosslink-distribution in the polymer. Only the relative exchange capacity of the copolymer products is affected. But it is specifically the homogeneity of the polymer-network that is important in defining ion-exchange resin performance. Resin characterics such as osmotic stability and resistance to fouling by organic substances depend on the attainment of a homogeneous structure. As a consequence, the efforts of researchers have in recent years been directed to a solution of this problem. A number of paths to the incorporation of homogeneity in ion-exchange resins has been examined. Schwachula et al.[110-114] have shown that the degree of heterogeneity of styrene - DVB copolymers is reduced by adding polar monomers such as acrylonitrile, ethylacrylate or vinylpyridine to the copolymerizing mixture. Copolymerization constant calculations for the styrene-p-DVB-acrylonitrile system ($r_{12} = 0.14$, $r_{13} = 0.41$, $r_{23} = 4.51$, $r_{21} = 0.50$, $r_{31} = 0.04$, $r_{32} = 0.204$)[115] are consistent with this observation. With these data a diagram, constructed to show the dependence of polymer composition on the monomer mixture composition demonstrates that as the amount of acrylonitrile increases the polymer composition approaches that of the azeotrope. Such three component copolymers are characterized by smaller differences in the swelling capacity of the central and outlying regions of the copolymer block[110]; their sulfonated exchanger products possess higher mechanical strength. The only undesirable feature of these resins is their polyfunctionality; the nitrile groups hydrolyze readily to form carboxyl groups[116].

A more uniformly crosslinked polystyrene matrix can be obtained for ion exchange resins by substituting diisopropenylbenzene (DPB) for DVB[117-121]. The styrene-DPB copolymerization constants are $r_1 = 1.2$, $r_2 = 0.8$ for m-DPB and $r_1 = 0.61$, $r_2 = 0.0$ for p-DPB[119]. These r values indicate that the styrene-DPB copolymer structure is not as irregularly crosslinked as the p-DVB-containing copolymers. Styrene-DPB copolymers have a higher swelling rate and swelling capacity than the styrene-DVB copolymers[122].

Ion-exchange resins obtained from the styrene-DPB copolymers are characterized by a high exchange and swelling-capacity and are quite permeable to organic ions[121]. The strongly basic anion-exchange resins sorb heparin and folic acid reversibly[117,120] while the sulfonated cation exchangers absorb up to 100-110 mg of insulin per gram of resin[121]. Ion-exchange resins containing iminodiacetic groups, are characterized by small changes in swelling capacity in different media[123].

An essentially different approach to the creation of copolymers which are more homogeneous has been through the introduction of additional crosslinking into the styrene-DVB copolymer product via side reactions with the chemical agent used to attach ionogenic groups. For example, it is known that the process of chloromethylation is accompanied by crosslinking side reactions as a result of the interaction of chloromethyl groups with unsubstituted phenyl rings of polystyrene[125]. Anderson, in 1964[124], attempted to correlate the dependence of properties of ion-exchange resins on the method of chloromethylation of styrene-DVB copolymers. He concluded that the formation of additional bonds in the crosslinked copolymer produced ion-exchange resins with high mechanical strength and small volume change in different eluents. Since the process of adding crosslinks to the styrene-DVB copolymer (especially in the case of a small DVB-content) must proceed statistically along the entire gel volume, the newly introduced bridges are more or less evenly distributed in the final polymer product. It is on this principle that the synthesis of isoporous ion-exchange resins has been based.

The catalysts employed for effecting the additional crosslinking of chloromethylated styrene copolymers are $AlCl_3$[126], Fe_2O_3[127], $SnCl_4$[129,171] or $ZnCl_2$[128]. Schwachula et al.[130] used a mixture of dimethyl formaldehyde, sulfuryl chloride and chlorosulfonic acid for carrying out simultaneous chloromethylation and crosslinking of the polystyrene. The free methyl chloride groups were then subjected to amination by standard techniques.

With this approach it was possible to synthesize the sulfonated cation exchangers as well, provided the sulfonation was carried out in the presence of formaldehyde with chlorosulfonic acid[131]. The cation exchangers so obtained are not distinguishable from ordinary resins by their chemical properties. Their mechanical strength depends on the initial polymer; the best results are obtained when the styrene copolymer containing 2% DVB and 2% acrylonitrile by weight, is subjected to additional crosslinking. Chloromethylated styrene-DVB copolymer can be additionally crosslinked with ethylene diamine (AB-22) or with hexamethylene diamine (AB-172). The further amination of these products with three-ethylamine produces polyfunctional isoporous ion-exchange resins[127].

Unfortunately, no methods have been established so far to analyze the uniformity of distribution of crosslinking bridges in the gel structure. An attempt [132] to make such an assessment of the isoporous ion-exchange resin structures by measuring their electrical conductivity was inconclusive.

At any rate the accessibility of their structure for amines that are large in size, provides evidence for the absence of densely crosslinked regions in the isoporous chloromethylated styrene copolymers (Table 6). Unlike isoporous structures, the degree of substitution of the methyl chloride groups in ordinary styrene-DVB copolymers drops sharply with an increase in size of the amine molecule.

The isoporous ion exchangers possess high osmtoic stability[2]. They do not become fouled by organic substances for an extended

TABLE 6

Degree of Amination (%) of Isoporous and Standard
Chloromethylated Styrene-DVB Copolymers[29]

	$N(CH_3)_3$	$N(CH_3)_2C_2H_4OH$	$NCH_3(C_2H_4OH)_2$	$N(C_2H_4OH)_3$
Isoporous resins	99	98	98	90
Standard resins	100	96	84	63

period of time. They are therefore widely used in demineralizing water[127].

E. Macronet Isoporous Polystyrene Structures and Ion-Exchange Resins Derived From Them

Of the ion exchange resins presently available, the most promising for use with organic substances are the macronet and isoporous types. One should expect to achieve the full potential of these two structures by combining their features in the polymer matrix of an ion-exchange resin. This objective cannot be obtained by the traditional monomer copolymerization reaction, but is approached by crosslinking the linear polymer chains (the polystyrene, in particular) in a solution or in a swollen state with different bifunctional compounds. The advantage of such a method for forming three-dimensional polymer structures is the following.

First, since by this approach the crosslinking agents are evenly distributed along the whole of the polymer solution (or gel) volume at the onset of the reaction, and since all the elementary polymer units are equivalent, the cross bridges that form in the course of the reaction should also be evenly distributed along the whole volume of the three-dimensional network produced. Such forming structures are related to the isoporous type.

Second, since the polymer matrix is formed in a solvated state and since the even distribution of the cross bridges minimizes the probability of occurrence of local strains in the ion-exchange resin grain during swelling, such an approach also contributes to enhancement of the osmotic stability of the ion-exchange resins.

In addition, this method of forming the polymer matrix permits the controlled variation over a wide range of properties of the polymers (and the ion-exchange resins obtained from them) by simply changing the amount or structure of the bridge-forming components; macronet and porous structures, in particular, are produced.

Any bifunctional compound capable of reacting with phenyl rings of polystyrene can be used as the agent for crosslinking the polystyrene. For example, bis-chloromethylated derivatives of the aromatic hydrocarbons with rigid elongated molecular structures such as 4,4'-bis-chloromethyldiphenyl (CMDP) and p-xylilene dichloride (XDC)[133], in particular, react with polystyrene according to the Friedel-Crafts reaction. In this case the bridge that is formed between the two polystyrene chains is two phenyl rings longer than the crosslinking agent:

$$\begin{array}{c}H_2C\\H-C\end{array}\!\!-\!\!\bigcirc\!\!-CH_2\left(\!\!-\!\!\bigcirc\!\!-\!\!\right)_n\!\!CH_2\!\!-\!\!\bigcirc\!\!-\!\!\begin{array}{c}CH_2\\CH\end{array}\quad n=1,2$$

One can use a strictly-defined amount of monochlorodimethyl ether MCDE as the linking agent in a similar manner. In this case the reaction proceeds through a polystyrene chloromethylation stage to produce structures with a known number of diphenylmethane bridges. By using the earlier methods of introducing additional crosslinks in the styrene-DVB polymer chloromethylation stage, it is difficult to monitor the degree of crosslinking in the end product. A small deviation from the standard synthesis conditions can lead to an important change in the gel structure and, consequently, to a change in the properties of the ion-exchange resin produced.

In all cases the dimensions of the crosslinking bridges significantly exceed those of the DVB molecule; therefore, the isoporous gels formed in this process are appropriately referred to as isoporous macronet structures.

INFLUENCE OF POLYMERIC MATRIX STRUCTURE ON PERFORMANCE

The hydrocarbon nature of the crosslinking bridges of the three-dimensional gel product leads to chemical stability comparable with that of the main polymeric chains. In addition, the bridges introduced into the polymer undergo chemical transformation like the main polystyrene chains during sulfonation, chloromethylation, nitration etc.

On crosslinking the dissolved polystyrene, the end product consists of irregularly shaped particles. In order to obtain granular macronet isoporous gels, the swollen granular styrene copolymer is used containing a small amount of any crosslinking agent.

Bis-chloromethyl derivatives of aromatic hydrocarbons possess a high reactivity, which allows the quantitative crosslinking of polymers [134], i.e. the preparation of product with a pre-determined degree of crosslinking, where the "degree of crosslinking" is defined as the ratio of the number of cross bridges to its sum with the number of non-substituted phenyl rings multiplied by 100. For example, if the reaction is conducted in the presence of 0.5 mol of crosslinking agent per mol of polystyrene expressed on a monomer basis, the degree of crosslinking of the end product is considered to be equal to 100%. However, the degree of crosslinking calculated in this fashion, at high concentration of the crosslinking agent, is somewhat arbitrary since the probability of interaction between the chloromethyl group of the crosslinking agent and the cross bridges already formed in the polymer increases. Therefore, even at a 100% degree of crosslinking, the polymer undoubtedly retains a portion of the non-substituted phenyl rings. This can be easily established by i.r. spectroscopy.

The physico-chemical properties of the macronet isoporous polystyrene gels were found to depend strongly on the manner of their production. The equilibrium swelling capacity of the macronet isoporous gels in toluene produced by crosslinking the dissolved polystyrene is considerably greater than that of the standard styrene-DVB copolymers with the same content of cross bridges [135]. However, the swelling capacity of macronet isoporous polystyrene structures

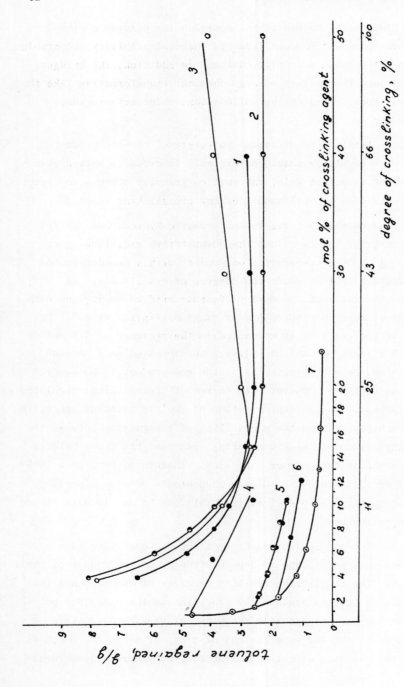

Fig. 2. Plot of equilibrium swelling capacity in toluene against the polymer crosslinking degree 1,2,3 - dissolved polystyrene (9.1 weight %) in dichlorethane crosslinked with CMDP(●) XDC (■) and MCDE (○). 4,6 - styrene-DVB copolymers (4-0.4 mol%, 5-1.4 mol%, 6-2.3 mol% of DVB) crosslinked with CMDP (●) and XDC (■) 7-styrene-DVB copolymers[26].

falls sharply when styrene-DVB copolymers are used as the starting material for further crosslinking (Fig. 2). In this case one can observe a distinct trend: the less DVB in the initial copolymer, the stronger it swells under conditions of further crosslinking (i.e. the Friedel-Crafts reaction) and the higher is the equilibrium swelling capacity value of the end product. The swelling capacity of these products though smaller than that of the polymers obtained by crosslinking polystyrene in solution, is always greater than the swelling capacity of standard styrene-DVB copolymers produced by copolymerization in the absence of a solvent.

These observations on the equilibrium swelling capacity of polymer product indicate convincingly the strong influence of dilution of the initial reaction mixture on the process of crosslink introduction.

This phenomenon is well-documented. As early as 1935 Staudinger[136] pointed out that the swelling capacity of polydivinylbenzene obtained in a benzene solution depended strongly on the initial monomer concentration. Mikes[137], in his study of styrene-DVB copolymers called attention to the fact that the swelling capacity as well as other polymer properties depended upon the degree of crosslinking and on the concentration of the initial monomeric mixture. Similar observations were made during study of the preparation of acrylamide gels[138], sephadexes[139], and glycerilmethacrylic polymers[140].

Very often two competing reactions are considered to account for the change in the swelling capacity of crosslinked polymers produced by copolymerization, as the concentration of the monomer mixture changes. Intrinsic molecular cyclization reactions that prevail in the more dilute solutions are presumed to become less important with increase in the concentration of the system while the interchain crosslinking reaction becomes the dominant factor bringing about a decline in the gel equilibrium swelling capacity. Dusek[141,142], Trostyanskaya[1,143], Vishnyakov[144], Determann[52] and others support this point of view. Davankov et al.[145] have offered another explanation of this phenomenon. In their opinion the three-dimensional network can be considered as a system of fragmented cells with separate units consisting of bridge-linked sites of different polymeric chains. Through these cells pass chains involved in the compoistion of other cells. In this case it is clear that the gel equilibrium swelling capacity is determined not only by the degree of polymer crosslinking, i.e. by the size of the separate cells, but also by the degree of interpenetration of these cells, which is determined by the conditions of their formation. The crosslinking of polymers in a solvating medium decreases the number of gel chains per unit of volume. It leads to structures with less interpenetration of the fragmented cells and to a greater equilibrium swelling capacity. Conversely the DVB crosslinking under conditions of a block monomer copolymerization produces gels with a maximal degree of interpenetration of the three-dimensional network cells, and with the least swelling capacity at the same degree of crosslinking. Thus, there is no unambiguous dependence between the gel equilibrium swelling capacity and its degree of crosslinking.

It has been regular procedure to attempt to characterize three-dimensional polymeric structures by their density through application of the Flory equation. A meaningful relationship between the equilibrium swelling capacity of crosslinked polymers and the mean molecular weight of the chain confined between two neighboring chemical crosslinks is considered to be established with this equation. A comparison of the density determined experimentally from the gel swelling capacity with that calculated theoretically from the quantity of bridge-forming components used is then the basis for certain conclusions. If the experimental density value exceeds appreciably the calculated value entanglement of polymeric chains is considered a likely explanation. When the calculated density is greater than the experimental value, this result is attributed to inefficiency of the bridge-forming components, with the crosslinking agent partially expended in forming the crosslinking bridges.

The failure of such analysis arises from the fact that a more complex relationship between swelling capacity and gel-crosslinking than that presumed by Flory in his theoretical treatment may exist. In the macronet isoporous systems, for example, the formation conditions of the three-dimensional network strongly influence swelling capacity even when the degree of crosslinking of the polymer remains constant. The inadequacy of the Flory treatment may be seen from consideration of the literature. Trostyanskaya et al.[146] found that with a styrene 2%-DVB polymer prediction was in agreement with experiment, i.e. all of the DVB was expended on the formation of crosslinking bridges, and physical entanglement of the polymer chains did not effect the copolymer swelling capacity. On the other hand, the experimentally determined network density of styrene copolymer containing 8% DVB was always significantly higher than the calculated one[147], even though, because of steric factors, a great portion of the bridge-forming elements (up to 30%) had participated in the copolymerization with only one double bond[148]. Discrepancy between theory and experiment was explained by the unsupported assumption that the degree of polymer chain entanglement increased with the increase in the number of crosslinks. Similarly, strong entanglement of the network was proposed since the theoretical density value of interpenetrating structures was considerably lower than the experimental density value determined from the swelling capacity. However, density calculations for the sulfonated exchangers obtained from these interpenetrating structures yield entirely different values, which to retain consistency in interpretation need to be explained by partial disentanglement of the networks [25]. Thus application of the well-known Flory equation can lead to an incorrect evaluation of the structural parameters of the three-dimensional polymer network, since this equation does not take into account all the important factors determining the swelling capacity of the crosslinked polymer.

Macronet isoporous polystyrene gels possess real porosity and a highly-developed inner surface at high degrees of crosslinking[149]. The inner surface area (S) increases with an increase in the number of crosslinking bridges introduced into the polymer and reaches

TABLE 7

Parameters of Porous Structures (S, m^2/g - inner specific surface area and W_o, cm^3/g - pore volume) of Polymers Crosslinked with 4,4'-Bis-Chloromethyldiphenyl (CMDP), p-Xylilenedichloride (XDC) and Monochlorodimethyl Ether (MCDE).

Crosslink-ing degree: %	CMDP		XDC		MCDE		porous polymers [51]	
	S	W_o	S	W_o	S	W_o	Name	S
17.7	0	-	0	-	0	-	Porapack Q	600
25	0	-	0	-	240	0.20	Porapack R	547
43	670	0.25	530	0.38	642	0.24	Porapack T	306
66	800	0.59	823	0.58	1000	0.35	Chromosorb 102	300
100	1009	0.63	956	0.62	990	0.62	Chromosorb 105	700

1000 m^2/g (Table 7). Simultaneously, the polymer pore volume (W_o) grows and reaches 0.63 cm^3/g for resins completely crosslinked. The type of crosslinking agent does not exert a noticeable influence on S and W_o.

However the concentration of the initial reaction medium strongly effects the value of S and W_o. With a decrease in the initial polystyrene solution concentration from 13.8 to 5.3 weight %, the S value of the network crosslinked with XDC (where the crosslinking degree equals 43%) decreases from 555 to 475 m^2/g, while the total pore volume, in this case, increases from 0.19 to 0.8 cm^3/g. Dilution of the initial reaction medium apparently provides an efficient means for controlling the structure of the gel product; the middle pore radius is affected in particular.

The right side of Table 7 lists the S values observed for macroporous styrene copolymers and their homologs with DVB. From these data it is apparent that the macronet isoporous polystyrene gels are superior to such familiar sorbents as Porapack Q and Chromosorb 105 on the basis of their inner surface area.

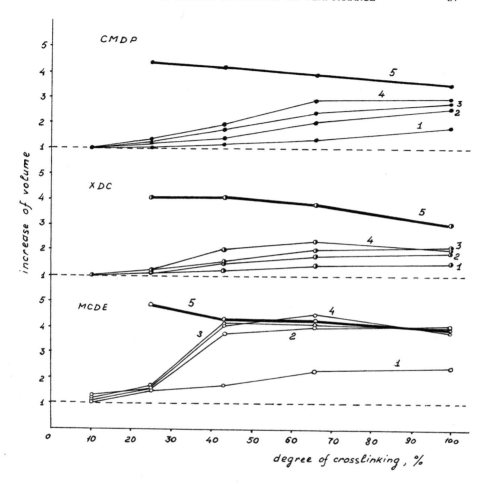

Fig. 3. Plot of volume swelling capacity of crosslinked polymers in different solvents against degree of crosslinking. Solvents: 1 - water; 2 - methanol; 3 - ethanol; 4 - hexane; 5 - toluene.

Unlike conventional styrene-DVB copolymers, macronet isoporous structures are capable of swelling not only in solvating solvents like toluene, but also in media which do not solvate the polystyrene[149]. This is illustrated in Figure 3 where the increase in volume of the polymers in toluene, hexane, ethanol, methanol and water with degree of crosslinking is plotted. (It should be noted that the polymers contain no functional groups bearing any affinity for the polar solvents.)

The ability of polymers crosslinked with CMDP and XDC to swell in non-solvating solvents, just perceptible at a 25% content of crosslinking bridges, increases with the further rise in crosslinking. The maximum increase in volume of these gels is in hexane; they swell to a smaller degree in ethanol, methanol and water. In all cases the difference in swelling ability in toluene and in non-solvating solvents decreases with a rise in the degree of crosslinking. The swelling of polymers crosslinked with MCDE in non-solvating media occurs at a lower degree of crosslinking (11%). Then it rises sharply and from a 43 to 100% crosslinking content, the volume swelling capacity of these gels in hexane, methanol and ethanol duplicates the swelling capacity value in toluene. Gels crosslinked with MCDE swell more in water than the others[†].

The explanation of this unexpected behavior of macronet isoporous polymers in non-solvating media needs to be sought in the structural pecularities of these gels. A three-dimensional isoporous polymer network forms when the polymer chains are in a solvated state. On removing the solvent from the gel that is formed, the polymer chains tend to become more densely packed. If the three-dimensional network contains a small amount of crosslinks, the polystyrene chains retain sufficient flexibility for this purpose. When the polymer crosslinking degree rises, their rigidity rises too, preventing the rearrangement of the polymer chains. On removing the solvent from a strongly crosslinked gel, part of the volume which is primarily occupied by the solvent, exists as pores that form in the polymer.

[†]The investigation of hydrophobic polymer swelling capacity in water is complicated by the fact that such behavior of the polystyrene structures cannot be measured directly. It is necessary to measure the swelling capacity in water by displacing methanol or ethanol from the swollen polymer. The value of the swelling capacity in water obtained in this way is independent of the particular solvent displaced. The measurement by this approach can be considered therefore, as an appropriate establishment of the gel-water equilibrium.

It is important to understand, however, that the tendency for contraction of strongly crosslinked gels leads to significant inner strain in the polymer chains. It is likely therefore that, in the general case, the framework formed in the presence of a solvent is strained in a non-solvated state. Even a weak interaction between the strained network and a solvent, for example, hexane is sufficient for part of the tension to be released at the expense of some increase in the gel volume. Perhaps, such relaxation of tension appears first in the framework crosslinked with monochlorodimethylether where the least elastic crosslinking bridges are formed. This results in the early appearance of inner surface area at low degrees of crosslinking and in a higher swelling capacity in poorly-solvating media.

It is necessary to point out once more that the internal strains existing in the dried isoporous gels are released when swelled in solvating media. It was under these conditions that the three-dimensional network of the resin was formed. Unlike standard styrene-DVB copolymers there are no locally strained polymer chains in such gels. The swelling pressure in structures with an even distribution of the crosslinking bridges produced in the presence of a solvating medium should be evenly distributed along the whole of the steric network.

The ability of the crosslinked gels to swell in non-solvating media is most probably a general property of these structures, their three-dimensional network forming in the presence of a solvent. It is mentioned twice in the literature that the macroporous styrene-DVB copolymers produced in a non-solvating medium such as n-heptane, swell in cyclohexane and increase their volume up to 40%[150,151]. The macroporous copolymers produced in the presence of toluene possess the same properties[29].

A distinguishing feature of the macronet isoporous styrene polymers is the ease of their transformation to ion-exchange resins. Even at a high content of crosslinking agent sulfonate groups are readily introduced by sulfonation under mild conditions[152]. The sulfonated exchangers that form possess an exchange capacity of from 5.0 to 5.4 meq/g when the degree of crosslinking of the polymer matrix is as high as 25%. The exchange capacity reduces to 4.2 to 4.7 meq/g when the degree of crosslinking is raised to 100%. The high capacity of the sulfonated macronet isoporous exchangers indicate that the crosslinking bridges do not act as ballast in the general polymer structure. They themselves can carry the same ionogenic groups as the main polystyrene chains. On the other hand, this observed behavior points to the great accessibility of the macronet isoporous gel structures to chemical reagents. This is likewise indicative of the facility with which other polymer transformations are carried out. Thus the hydrolysis of methyl esters of α-amino acids linked to the chloromethylated macronet isoporous polymer (6% crosslinking) proceeds considerably easier than the hydrolysis of esters linked to standard resins containing only 0.8% of DVB[153]. Macronet styrene copolymers are extremely useful as activated polymeric esters for peptide synthesis[154,155].

The swelling capacity of sulfonated macronet isoporous exchangers in water is significantly higher than that of standard resins with the same amount of crosslinking agent. Even when highly crosslinked from 66 to 100% they can absorb as much water as conventional cation exchangers with 1-2% DVB.

A comparison has been made of the permeability of sulfonated ion-exchange resins obtained from the macronet isoporous polystyrene gels with those obtained from styrene-DVB copolymers at the same swelling capacity in water. Fig. 4 illustrates the time of half-exchange ($\tau_{0.5}$) of hydrogen ions for tetrabutylammonium ions as a function of the swelling capacity of the sulfonated resins in water. With a decrease in the swelling capacity of the cation exchangers in water, $\tau_{0.5}$ rises. The process, however, proceeds very rapidly even for resins with the least swelling capacity, $\tau_{0.5}$ not exceeding 1.5 minutes. All the experimental points, as is seen, correlate with the swelling capacity of the resins in water; there is no correlation with the degree of crosslinking. All points obtained for the sulfon-

INFLUENCE OF POLYMERIC MATRIX STRUCTURE ON PERFORMANCE 71

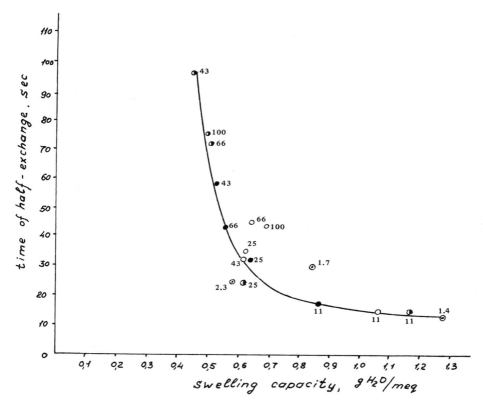

Fig. 4. Plot of time of half-exchange H^+ - $(C_4H_9)_4N^+$ against the swelling capacity of sulfonated exchangers in water. Polymers crosslinked with CMDP (●), XDC (◐), MCDE (o) and DVB (⊙). The numbers by each point denote the crosslinking degree of the exchangers.

ated resins crosslinked with CMDP and XDC fall on one curve. The ion-exchange process is somewhat slower for resins crosslinked with MCDE at a crosslinking content exceeding 25% (66 and 100%). Points obtained with the standard sulfonated exchangers seem to fall on or near this curve as well. The data obtained seem to indicate the absence of any significant influence of the polymeric matrix structure of the cation exchanger on the kinetics of the process. Apparently, the ion-exchange rate at the initial stage is determined largely by the content of water in the cation exchanger, (not by the crosslinking degree of the resins, the length of the crosslinking bridges or the nature of distribution of the crosslinks).

However, the time for full saturation of the resin by tetrabutylammonium is several times longer for the macronet cation exchangers (with the degree of crosslinking of the polymer matrix ranging from 43 to 100%) than for the standard lower crosslinked resins with the same swelling capacity in water. This result is quite reasonable. The high network elasticity of copolymers containing 1-2% DVB facilitates the penetration of ions into the sterically hampered sorption centers. Conversely, the small deformation of polymer network cells that is possible in the rigid matrices of the highly crosslinked macronet isoporous resins cannot significantly facilitate the diffusion of large ions to the sterically hampered sites.

One exceptionally important feature of the sulfonated macronet isoporous resins is their unexpectedly high sorption capacity for organic ions as large as tetrabutylammonium. These exchangers irrespective of the type of crosslinking agent and at any crosslinking degree, can be fully saturated by tetrabutylammonium ion. In the conventional sulfonated polystyrene cation exchangers a rise in the degree of crosslinking leads to a sharp decrease in the exchange capacity for tetrabutylammonium ion; only half of the ionogenic groups are accessible for this ion at a 10% DVB content (Table 1).

In addition to their high permeability for large organic ions the highly crosslinked (>30-40%) macronet isoporous cation exchangers are characterized by small volume changes upon replacing the eluent. In the H^+-form their swelling capacity (0.4-0.7 g of H_2O/meq is the same in water, 0.1 N HCl and 1N HCl. When converted to the K^+ ion form in 0.1N KOH there is no change in volume as well. The volume of these resins is insensitive to the counter-ion and the ionic strength of electrolyte solution in contact with it. With the H^+-ion form of the lower crosslinked standard styrene-DVB cation exchangers which exhibit a comparable swelling capacity in water (0.58g of H_2O/meq, 2.3% of DVB) there is an 18-20% decrease in volume under comparable conditions.

When replacing water for ethanol or dioxane, the macronet isoporous resin volume remains practically the same, while with the lower-crosslinked standard resins the decrease in volume in these solvents is 12 and 50%, respectively. Even in a non-polar solvent such as benzene, the macronet isoporous resins decreases in volume only by 15-20%, whereas the conventional resins do not swell in benzene at all. Macronet isoporous resins behave like oleophilic resins and their swelling behavior is unaffected by media of different polarity[157-163]. When the degree of crosslinking is less than 40% the volume of the macronet isoporous sulfocation exchangers does of course vary to some extent with medium changes.

The absence of volume change in strongly-crosslinked macronet isoporous resins is explained by the very same factors which leads to the swelling of polymer matrices in poorly-solvating media, namely, by the high rigidity of the polymeric matrix.

Thus, the introduction of sufficiently large amounts of long rigid bridges into the polymer by crosslinking of the linear polystyrene chains in solution leads to the formation of structures which, on the one hand, maintain a high permeability for large organic ions and, on the other hand, provide a sizeable resistance to volume change in different eluents. Structures simultaneously combining both of these paroperties are not obtained by the traditional monomer copolymerization method.

III. CONCLUSION

The continuing search for new methods to synthesize ion-exchange resins and thereby improve their performance characteristics has been and continues to be quite successful.

REFERENCES

1. G.V. Samsonov, E.B. Trostyanskaya, G.E. Yelkin, "Ion Exchange. Sorption of Organic Substances," Nauka, Leningrad, 1969, p. 7.
2. R.M. Wheaton, M.J. Hatch, "Ion Exchange", ed. by J. Marinsky, Marcel Dekker, Inc., New York, 1969, Vol. 2, p. 191.
3. G.V. Samsonov, V.A. Pasechnik, Usp. Khimii. 38, 1257 (1969).
4. R.H. Wiley, E.E. Sale, J. Polymer Sci., 42, 491 (1960).
5. R.H. Wiley, B. Davis, J. Polymer Sci., B1, 463 (1963).
6. R.H. Wiley, W.K. Mathews, K.F. O'Driscoll, J. Macromol. Sci., A1, 503 (1967).
7. R.H. Wiley, Rao S. Prabhakara, J-I. Jin, K.S. Kim, J. Macromol. Sci., A4, 1453 (1970).
8. J. Malinský, J. Klaban, K. Dušek, Coll. Czech. Chem. Comm., 34, 711 (1969).
9. J. Malinský, J. Klaban, K. Dušek, J. Macromol. Sci., A5, 1071 (1971).
10. S.B. Makarova, Zh.M. Litvak, I.A. Vakhtina, E.V. Yegorov, Vysokomol. Soyed., 13A, 2160 (1971).
11. R.H. Wiley, J-I. Jin, E. Reich, J. Macromol. Sci., A4, 341 (1970).
12. V.D. Yenalyev, A.A. Kondratovich, E.P. Gendrikov, G.S. Dedovets, Plast. Massy, N8, 5 (1965).
13. S.I. Belfer, Thesis, NIIPM, Moscow, 1970.
14. R.H. Wiley, Chim. et Ind., 94, 602 (1965).
15. S.I. Belfer, G.K. Saldadze, E.A. Kudryavtsev, S.A. Stepanyan, A.N. Shabadash, Plast, Massy, N8, 24 (1970).
16. R.H. Wiley, J. Polymer Sci., A1, 1892 (1966).
17. S.B. Makarova, T.A. Aptova, N.M. Vinogradova, T.M. Chernyavskaya, E.V. Yegorov, Plast. Massy, N10, 5 (1970).
18. R.H. Wiley, J.K. Allen, S.P. Chang, K.E. Musselman, T.K. Venkatachalam, J. Phys. Chem., 68, 1776 (1964).
19. E.B. Trostyanskaya, S.B. Makarova, T.A. Aptova, N.M. Vinogradova, Vysokomol. Soyed., 9A, 1066 (1967).
20. S.I. Belfer, K.M. Saldadze, E.G. Gintsberg, B.M. Kovarskaya, Zh. Fix. Khim., 44, 1104 (1970).
21. T.K. Brutskus, K.M. Saldadze, E.A. Uvarova, M.A. Fedtsova, S.I. Belfer, Zh. Fix. Khim., 47, 1528 (1973).
22. K.M. Saldadze, Ts.S. Kurtskhaliya, Past. Massy, N6, 5 (1968).
23. R.H. Wiley, T.K. Venkatachalam, J. Polymer Sci., A3, 1063 (1965).

INFLUENCE OF POLYMERIC MATRIX STRUCTURE ON PERFORMANCE

24. R.M. Bakayeva, G.V. Samsonov, Coll. "Selective Ion-Exchange Sorption of Antibiotics", Trans. of the Leningrad Chemico-Pharmachological Institute, Issue 25, Leningrad, 1968, p.63.
25. J.R. Millar, J. Chem. Soc., 1311 (1960).
26. T.R.E. Kressman, J.B. Millar, Chem. and Ind., 1833 (1961).
27. J.R. Millar, D.G. Smith, W.E. Marr, J. Chem. Soc., 1789 (1962).
28. T.R.E. Kressman, "Ion Exchange in the Process Industries", Soc. Chem. Ind., London, 1969, P. 3.
29. J.R. Millar, D.G. Smith, W.E. Marr, T.R.E. Kressman, J. Chem. Soc., 218 (1963).
30. W.D. Lloyd, T.Ir. Alfrey, J. Polymer Sci., $\underline{62}$, 301 (1962).
31. R.H. Wiley, J.T. Badcett, J. Macromol. Sci., $\underline{A2}$, 103 (1968).
32. J.B. Millar, D.G. Smith, W.E. Marr, T.R.E. Kressman, J. Chem. Soc., 2779 (1963).
33. Teidzi Tsuruta, "Reactions of Synthetic Polymers Production", Goskhimizdat, Moscow, 1963, p. 42.
34. E.B. Trostyanskaya, S.B. Makarova, T.A. Aptova, I.N. Murashko, Vysokomol. Soyed., $\underline{7}$, 2083 (1965).
35. E.B. Trostyanskaya, A.S. Tevlina, F.A. Naumova, Vysokomol. Soyed., $\underline{5}$, 1240 (1963).
36. A.S. Tevlina, G.S. Kolesnikov, G.V. Samsonov, L.H. Dmitriyanko, A.E. Chuchin, "Chemically Active Polymers and their Application", Khimiya, Leningrad, 1969, p. 24.
37. A.S. Tevlina, S.F. Sadova, Zh. Prikl. Khim., $\underline{38}$, 1643 (1965).
38. H. Small, J. Am. Chem. Soc., $\underline{90}$, 2217 (1968).
39. A.Sh. Genedi, G.V. Samsonov, Coll. "Selective Ion-Exchange of Antibiotics", Transactions of the Leningrad Chemico-Pharmacological Institute, Issue 25, Leningrad, 1968, p. 164.
40. E.I. Denisenko, B.N. Trushin, "Chemistry and Chemical Technology", Coll. Scientific Papers N36, Kuzbass Polytechnical Institute, Kemerovo, 1972, p. 141.
41. K.M. Saldadze, S.I. Belfer, S.M. Tkachuk, Ts.S. Kurtskhaliya, Plast. Massy, $\underline{N10}$, 6 (1968).
42. P.S. Belov, S.I. Belfer, I.I. Iwanova, K.M. Saldadze, Zh. Prikl. Khim., $\underline{46}$, 2031 (1973).
43. J.A. Mikes, "Ion Exchange in the Process Industries", Soc. Chem. Ind., London, 1969, P. 16.
44. K.A. Kun, R. Kunin, J. Polymer Sci., $\underline{B2}$, 587 (1964).
45. R. Kunin, "Ion Exchange in the Process Industries", London, 1969, p. 10.
46. J.R. Millar, D.G. Smith, T.R.E. Krussman, J. Chem. Soc., 304 (1965).

47. R.R. Yergozhin, "Monomers and Polymers", Trans. of the Institute of Chemical Sciences, Academy of Sciences, KazSSR, Alma-Ata-Vol.23, part 2, 1969, p.53.
48. J. Seidl, J. Malinský, K. Dušek, W. Heitz, Adv. Polymer Sci., 5, 113 (1967).
49. J. Seidl, J. Malinský, K. Dušek, Plast. Massy, N12, 7 (1963).
50. K.I. Sakodynsky, L.I. Mosyava, "Gas Chromatography", NIITEKhim, Moscow, Issue 7, 1967, p.18.
51. K.I. Sakodynsky, L.I. Panina, Zh. Anal. Khim., 27, 1024 (1972).
52. H. Determan, "Gelchromatographie", Springer-Verlag, Berlin-Heidelberg-New York, 1967.
53. V.V. Alekseyeva, Ju.I. Ostroushko, L.I. Vodolazov, Ju.N. Fedulov, G.S. Pakhomova, T.A. Tirman, Zh. Prikl. Khim., 42, 396 (1969).
54. R. Kunin, E.F. Meitzner, J.A. Oline, S.A. Fisher, N. Frisch, Ind. Eng. Chem., Prod. Res. Dev. 1, 140 (1962).
55. R. Kunin, E.F. Meitzner, N. Bortnik, J. Am. Chem. Soc., 84, 305 (1962).
56. T.K. Brutskus, K.M. Saldadze, E.A. Uvarova, E.I. Lyustgarten, Zh. Fis. Khim., 47, 353 (1973).
57. K.M. Saldadze, T.K. Brutskus, M.A. Fedtsova, E.A. Uvarova, E.I. Lyustgarten, T.S. Semyenova, Zh. Fiz. Khim., 44, 2815 (1970).
58. K. Dušek, Coll. Czech. Chem. Comm., 30, 3804 (1965).
59. G.E. Yelkin, S.F. Belaya, V.A. Dinsburg, G.B. Samsonov, All Union Inter-College Conference on Chromatography. Brief reports and reviews of papers, Voronezh, 1965, p. 34.
60. G.V. Samconov, V.A. Dinaburg, K.M. Genender, G.E. Yelkin, S.F. Belaya, V.S. Jurchenko, Abstracts of Papers. 12 Conference of the Institute of High-Molecular Compounds, Academy of Sciences, USSR, Leningrad, 1965.
61. V.A. Dinaburg, O.P. Kolomeitsev, K.M Genender, K.B. Musabekov, A.A. Vansheidt, Abstracts of Papers. 13 Conference of the Institute of High Molecular Compounds, Academy of Sciences, USSR, Leningrad, 1966.
62. K.B. Musabekov, V.A. Pasechnik, A.Sh. Genedia, G.E. Yelkin, V.A. Dinaburg, G.V. Samsonov, Abstracts of Papers. 14 Conference of the Institute of High Molecular Compounds, Academy of Sciences, USSR, Leningrad, 1967.
63. G.V. Samsonov, V.A. Dinaburg, S.F. Belaya, V.A. Pasechnik, G.E. Yelkin, V.S. Yurchenko, K.M. Genender, Zh. Prikl. Khim., 44, 859 (1971).
64. K.B. Musabekov, V.A. Dinaburg, G.V. Samsonov, Zh. Prikl. Khim. 42, 82 (1969).

65. N.N. Kuznetsova, K.P. Papukova, A.N. Libel, G.V. Samsonov, Patent USSR N 290031,"Discoveries. Inventions. Industrial Designs. Trade Marks", N2, 60 (1971).
66. O.P. Kolomeitsev, N.N. Kusnetsova, V.A. Dinaburg, Coll. "Ion Exchange and Ion Exchange Resins," Nauka, Leningrad, 1970, p.48.
67. V.A. Dinaburg, G.V. Samsonov, K.M. Genender, V.A. Pasechnik, V.S. Yurchenko, G.E. Yelkin, S.F. Belaya, Zh. Prikl. Khim., 41, 891 (1968).
68. O.P. Kolomeitsev, N.N. Kuznetsova, Zh. Prikl. Khim., 45, 1978 (1972).
69. S.F. Belaya, G.E. Yelkin, G.V. Samsonov, Coll. "Selective Ion Exchange Sorption of Antibiotics". Transactions of the Leningrad Chemico - Pharmacological Institute, Issue 25, Leningrad, 1968 p. 147.
70. K.B. Musabekov, V.A. Pasechnik, G.V. Samsonov, Zh. Fis. Khim., 44, 991 (1970).
71. K.B. Musabekov, V.A. Pasechnik, Vestnik Academy of Sciences KazSSR, N5, 54 (1970).
72. K.B. Musabekov, V.A. Pasechnik, G.V. Samsonov, Isvestiyel, Academy of Sciences, KazSSR, Chemical Series, N5, 64 (1968).
73. V.A. Pasechnik, K.B. Musabekov, G.V. Samsonov, Coll. "Synthesis, Structure and Properties of Polymers", Transactions of the 15 Conference of the Institute of High Molecular Compounds, Academy of Sciences, USSR, Leningrad, 1970, p. 265.
74. V.S. Yurchenko, G.I. Kilfin, K.B. Musabekov, V.A. Pasechnik, G.I. Samsonov, Coll. "Selective Ion Exchange Sorption of Antibiotics". Transactions of the Leningrad Chemico-Pharmacological Institute, Issue 25, 1968, p. 121.
75. K.B. Musabekov, E.E. Yergozhin, L.P. Shapovalova, Isvestiyel, Academy of Sciences, KazSSR, Chemical Series, N6, 59 (1972).
76. S.F. Belaya, G.E. Yelkin, G.V. Samsonov, Kolloid Zhurnal, 33, 645 (1971).
77. B.V. Moskvichev, V.S. Yurchenko, A.Ch. Genedi, B.Sh. Chokina, G.V. Samsonov, "Synthesis, Structure and Properties of Polymers", Transactions of the 15 Conference of the Institute of High Molecular Compounds, Academy of Sciences, Nauka, Leningrad, 1970, p. 263.
78. A.Sh. Genedi, G.V. Samsonov, Zh. Fiz. Khim., 44, 3128 (1970).
79. G.V. Samsonov, B.V. Moskvichev, V.S. Yurchenko, A.Sh. Genedi, B.Sh. Chokina, Coll. "Ion Exchange and Ion Exchange Resins", Nauka, Leningrad, 1970, p. 142.
80. A.Sh. Genedi, B.V. Moskvichev, G.V. Samsonov, Zh. Prikl. Khim., 43, 1171 (1970).

81. G.V. Samsonov, E.B. Trostyanskaya, G.E. Yelkin, "Ion Exchange. Sorption of Organic Substances", Nauka, Leningrad, 1969, p. 224.
82. G.V. Samsonov, A.Sh. Genedi, K.B. Musabekov, G.E. Yelkin, Zh. Prikl. Khim., $\underline{41}$, 1540 (1968).
83. A.Sh. Genedi, G.V. Samsonov, Kolloid Zhurnal, $\underline{31}$, 674 (1969).
84. K.B. Musabekov, V.S. Yurchenko, G.V. Samsonov, Isvestiyel, Academy of Sciences, KazSSR, Chemical Series, $\underline{N2}$, 58 (1969).
85. B. Trémillon, "Les Separations par Les Résines Ecchangeuses d'ions", Gauthier-Vellars-Paris, 1965.
86. O.P. Kolomeitsev, N.N. Kuznetsova, Vysokomol. Soyed., $\underline{14B}$, 500 (1972)
87. O.P. Kolomeitsev, N.N. Kuznetsova, Vysokomol. Soyed., $\underline{13A}$, 1899 (1971).
88. O.P. Kolomeitsev, V.A. Pasechnik, N.N. Kuznetsova, G.V. Samsonov, Vysokomol. Soyed., $\underline{14A}$, 1746 (1972).
89. G.S. Kolesnikov, Vysokomol. Soyed., $\underline{6}$, 559 (1964).
90. E.E. Yergozhin, M. Kurmanaliyev, "Monomers and Polymers", Transactions of the Institute of Chemical Sciences, Academy of Sciences, KazSSR, Alma-Ata, $\underline{V32}$, part 6, 1972, p.34.
91. E.E. Yergozhin, B.A. Zhubanov, Yu.A. Kushnikov, L.N. Prodius, Isvestiyel, Academy of Sciences, KazSSR, Chemical Series, $\underline{N3}$, 44 (1970).
92. M. Kurmanaliyev, E.E. Yergozhin, S.R. Rafikov, Isvestiyel, Academy of Sciences, KazSSR, Chemical Series, $\underline{N3}$, 49 (1972).
93. M. Kurmanaliyev, E.E. Yergozhin, Vestnik, Academy of Sciences, KazSSR, $\underline{N9}$, 53 (1971).
94. M. Kurmanaliyev, E.E. Yergozhin, Isvestiyel, Academy of Sciences, KazSSR, Chemical Series, $\underline{N5}$, 67 (1972).
95. E.E. Yergozhin, L.N. Prodius, S.R. Rafikov, Isvestiyel, Academy of Sciences, KazSSR, Chemical Series, $\underline{N2}$, 53 (1972).
96. V.N. Prusova, E.E. Yergozhin, B.A. Zhubanov, S.R. Rafikov, Isvestiyel, Academy of Sciences, KazSSR, Chemical Series, $\underline{N2}$, 59 (1972).
97. E.E. Yergozhin, S.R. Rafikov, B.A. Zhubanov, L.N. Prodius, Isvestiyel, Academy of Sciences, USSR, Chemical Series, 2128 (1970).
98. E.E. Yergozhin, B.A. Zhubanov, V.N. Prusova, S.R. Rafikov, Isvestiyel, Academy of Sciences, USSR, Chemical Series, 972 (1972).
99. B.N. Trushin, Thesis, D.I. Mendeleev Chemical Engineering Institute, Moscow, 1968.

100. B.N. Trushin, A.B. Davankov, V.V. Korshak, Vysokomol. Soyed., 9A, 1140 (1967).

101. B.N. Trushin, M.I. Lobova, "Chemistry and Chemical Technology". Coll. Scientific Papers N26, Kuzbass Polytechnical Institute, Kemerovo, 1970, p. 233.

102. B.N. Trushin, A.B. Davankov, V.V. Korshak, Coll. "Chemistry and Technology of Organic Substances and High Molecular Compounds". Transactions of the D.I. Mendeleev Institute of Chemical Engineering, Issue 57, Moscow, 1968, p. 119.

103. B.N. Trushin, A.B. Davankov, V.V. Korshak, ibid., 1968, p.124.

104. B.N. Trushin, A.B. Davankov, "Chemistry and Chemical Technology", Coll. Scientific Papers N26, Kuzbass Polytechnical Institute, Kemerovo, 1970, p. 226.

105. B.N. Trushin, A.B. Davankov, V.M. Laufer, A.B. Belousova, Zh. Prikl. Khim., 41, 1293 (1968).

106. V.N. Prusova, E.E. Yergozhin, Isvestiyel, Academy of Sciences, KazSSR, Chemical Series, N5, 62 (1972).

107. E.E. Yergozhin, B.A. Zhubanov, Yu.A. Kushnikov, V.N. Prusova, Isvestiyel, Academy of Sciences, KazSSR, Chemical Series, N1, 44 (1972).

108. E.E. Yergozhin, B.A. Zhubanov, L.N. Prodius, K.B. Musabekov, Z.A. Nurkhodzhayeva, Isvestiyel, Academy of Sciences, KazSSR, Chemical Series, N5, 54 (1971).

109. V.N. Prusova, E.E. Yergozhin, Isvestiyel, Academy of Sciences, KazSSR, Chemical Series, N1, 51 (1972).

110. G. Schwachula, F. Wolf, Plaste u. Kautschuk, 14, 879 (1967).

111. G. Schwachula, F. Wolf, Plaste u. Kautschuk, 15, 33 (1968).

112. G. Schwachula, F. Wolf, H. Schmidt, Plaste u. Kautschuk, 14, 802 (1967).

113. G. Schwachula, F. Wolf, H. Katche, Plaste u. Kautschuk, 19, 731 (1972).

114. G. Schwachula, H. Schmidt, Plaste u. Kautschuk, 18, 577 (1971).

115. G. Schwachula, F. Wolf, H. Gatzmanga, Plaste u. Kautschuk, 17, 255 (1970).

116. H. Hoffmann, G. Schwachula, Plaste u. Kautschuk, 18, 98 (1971).

117. A.S. Tevlina, V.V. Korshak, L.E. Frumin, Yu.V. Kamnyev, "Production and Processing of Plastics and Synthetic Resins," Coll. Tech. and Econ. Information, N7, NIIPM, Moscow, 1973, p. 8.

118. G.S. Kolesnikov, A.S. Tevlina, L.E. Frumin, A.I. Kirilin, Vysokomol. Soyed., 13A, 549 (1971).

119. V.V. Korshak, A.S. Tevlina, L.E. Frumin, A.I. Kirilin, T.D. Dedkova, T.N. Ivanov, Vysokomol. Soyed., 14B, 394 (1972).

120. S.I. Shvarts, G.P. Zhivotova, Yu.A. Ivanov, L.E. Frumin, A.S. Tevlina, "Hormone and Organotherapy Preparations in Medicine" VNIIA, Moscow, 1971, p. 350.

121. L.E. Frumin, Thesis, D.I. Mendeleev Chemical Engineering Institute, Moscow, 1973.

122. E.I. Lyustgarten, A.B. Pashkov, S.M. Tkachuk, T.K. Brutskus, S.L. Kreinina, T.V. Dyachenko, B.S. Khmelnitskaya, "Production and Processing of Plastics and Synthetic Resins", Coll. Tech. and Econ. Information, N7, NIIPM, Moscow, 1973, p. 7.

123. R. Hering, Z. Chemie, 5, 149 (1965).

124. R.E. Anderson, Ind. Eng. Chem., Prod. Res. Dev., 3, 85 (1964).

125. G.D. Jones, Ind. Eng. Chem., 44, 2686 (1952).

126. U.S. Patent 2635061 (1953); C.A., 47, 6164h (1953).

127. A.B. Pashkov, M.I. Itkina, E.I. Lyustgarten, et. al. "Production and Processing of Plastics and Synthetic Resins", Coll. Tech. and Econ. Information, N7, NIIPM, Moscow, 1973, p. 2.

128. B.N. Trushin, ibid., p. 3.

129. B.N. Trushin, V.L. Veldyaskina, "Chemistry and Chemical Technology", Coll. Scientific Papers, N36, Kuzbass Polytechnical Institute, Kemerovo, 1971, p. 139.

130. R. Hauptmann, G. Schwachula, Z. Chem., 8, 227 (1968).

131. G. Schwachula, R. Hauptmann, I. Kain, 13 IUPAC Microsymposium on Macromolecules, Prague, 1973, Abstracts, C3.

132. V.D. Grebenyuk, Zh. Fiz. Khim., 44, 3149 (1970).

133. V.A. Davankov, S.V. Rogozhin, M.P. Tsyurupa, U.S. Patent 3729457, C.A., 75, 6841v (1971).

134. V.A. Davankov, S.V. Rogozhin, M.P. Tsyurupa, Vysokomol. Soyed., 15B, 463 (1973).

135. M.P. Tsyurupa, S.V. Rogozhin, V.A. Davankov, "Production and Processing of Plastics and Synthetic Resins", Coll. Tech. and Econ. Information N7, NIIPM, Moscow, 1973, p. 3.

136. H. Staudinger, E. Husemann, Chem. Ber., 68B, 1618 (1935).

137. J.A. Mikes, J. Polymer Sci., 30, 615 (1958).

138. S. Hjertén, Arch. Biochem. Biophys., 1, 147 (1962).

139. P. Flodin, "Dextran Gels and Their Application in Gel Filtration Upsala," 1962.

140. M.F. Refojo, J. Appl. Polymer Sci., 9, 3161 (1965).

141. K. Dušek, J. Malinský, Chem. Průmysl., 16 (41), 219 (1966).

142. K. Dušek, Coll. Czech. Chem. Comm., 32, 1182 (1967).
143. E.B. Trostyanskaya, P.G. Babaevsky, Usp. Khimii, 40, 117 (1971).
144. I.I. Vishnyakov, Vysokomol. Soyed., 7, 239 (1965).
145. V.A. Davankov, S.V. Rogozhin, M.P. Tsyurupa, Angew. Macromol. Chem., 32, 145 (1973).
146. E.B. Trostyanskaya, S.B. Makarova, T.A. Aptova, I.N. Murashko, E.V. Yegorov, Coll. "Synthesis and Properties of Ion Exchange Materials", Nauka , Moscow, 1968, p. 17.
147. K.M. Saldadze, S.I. Belfer, Plast. Massy., N3, 10 (1967).
148. K. Dušek, Coll. Czech. Chem. Comm., 27, 2841 (1962).
149. V.A. Davankov, S.V. Rogozhin, M.P. Tsyurupa, 13th IUPAC Microsymposium on Macromolecules, Prague, 1973, Abstracts, C2.
150. E.I. Lyustgarten, B.P. Li, A.B. Pashkov, N.B. Skakalskaya, T.I. Davydova, M.A. Zhukova, Plast. Massy, N5, 7 (1964).
151. E.I. Lyustgarten, A.B. Pashkov, T.I. Davydova, N.B. Skakalskaya, Coll. "Chemicallly Active Polymers and Their Application", Khimiya , Leningrad, 1969, p. 28.
152. M.P. Tsyurupa, S.V. Rogozhin, V.A. Davankov, 13th IUPAC Microsymposium on Macromolecules, Prague, 1973, Abstracts, E5.
153. I. Peslyakas, Thesis, Vilnius, 1972.
154. S.V. Rogozhin, Yu.A. Davidovich, S.M. Andreyav, A.P. Yurtanov, Dokl. Akad. Nauk, USSR, 211, 1356 (1973).
155. S.V. Rogozhin, Yu.A. Davidovich, S.M. Andreyev, A.P. Yurtanov, Dokl. Akad. Nauk., USSR, 212, 108 (1973).
156. H.P. Gregor, P. Teyssie, G.K. Hoeschele, R. Feinland, M. Shida, A. Tsuk, Polymer Preprints, 5, 873 (1964).
157. A.G. Tsuk, H.P. Gregor, Polymer Preprints, 5, 878 (1964).
158. A.G. Tsuk, H.P. Gregor, J. Am. Chem. Soc., 87, 5534 (1965).
159. H.P. Gregor, G.K. Hoeschele, J. Patenza, A.G. Tsuk, R. Feinland, M. Shida, P. Teyssie , J. Am. Chem. Soc., 87, 5525 (1965).
160. Mitsuyzo Shida, H.P. Gregor, J. Polymer Sci., 4A, 1113 (1966).
161. H.P. Gregor, "Ion Exchange in the Process Industries", Soc. Chem. Ind., London, 1969, P. 436.
162. J. Haklits, J. Szanto, Plaste u. Kautschuk, 18, 175 (1971).
163. S.B. Makarova, E.M. Pakhomova, O.V. Babina, E.V. Yegorov, Plast. Massy., N8, 15 (1969).
164. T. Yamashita, Bull. Chem. Soc. Japan, 45, 195 (1972).

Chapter 3

SPECTROSCOPIC STUDIES OF ION EXCHANGERS

Carla Heitner-Wirguin

Department of Inorganic & Analytical Chemistry
Hebrew University, Jerusalem
Israel

I.	INTRODUCTION	84
II.	PREPARATION OF SAMPLES FOR SPECTRAL MEASUREMENTS AND THE EVALUATION OF QUANTITATIVE PARAMETERS	84
	A. Near Infrared, Visible and Ultraviolet Spectra	84
	B. Infrared Spectra	87
	C. Raman Spectra	87
	D. Other Methods Used	89
III.	SPECTROSCOPIC STUDIES ON ZEOLITES	89
	A. General Properties and Uses	89
	B. Structure of the Framework	91
	C. Water and Hydroxyl Groups in Zeolites	98
	D. Ion Exchange and Catalytic Properties	101
	1. Manganese	101
	2. Iron	103
	3. Cobalt	106
	4. Nickel	110
	5. Copper	110
	E. Concluding Remarks	114
IV.	ORGANIC ION EXCHANGERS	116
	A. General Properties	116
	B. Complex Species Sorbed on Cation and Anion Exchange Resins	120
	C. Complex Species Sorbed on Chelating Ion Exchangers	141
	D. Spectra of Ion Exchange Membranes	148
	E. Concluding Remarks	152
V.	ADDENDUM	154
	REFERENCES	157

I. INTRODUCTION

The ion exchange process has been studied for many years. As a result of the tremendous amount of research performed, it is today quite well understood and many industrial applications have been developed. Despite large differences existing between the various types of ion exchangers, there is one structural property which is common to all of them: a polymeric framework which makes them insoluble in water. This framework has proved in recent years to be a suitable matrix for spectroscopic investigation. From the spectroscopic examination of various ion exchangers, much information has been obtained for a) the elucidation of their structure and for b) the identification of the ion species sorbed.

It is the purpose of this chapter to survey, first some recent advances made through spectral studies in the elucidation of the structure and properties of zeolites and, second, to show that the use of these techniques provides important information on the nature of complex species through their sorption on organic ion exchangers.

II. PREPARATION OF SAMPLES FOR SPECTRAL MEASUREMENTS AND THE EVALUATION OF QUANTITATIVE PARAMETERS

As ion exchangers are, in general, solids of the crystalline or gel type, the classical techniques used for liquids in spectroscopy can very seldom be used. Special preparation of samples must usually be made in order to obtain spectra sufficiently resolved for interpretation.

A. Near Infrared, Visible and Ultraviolet Spectra

The techniques for measuring absorption spectra in these solids in the NIR, visible and UV regions vary from author to author since many difficulties are encountered. It is quite understandable that many of the early studies were made by measuring reflectance spectra since they are much easier to accomplish with solids. Such measurements were made, for example, by Rutner (1) on chlorocomplexes of cobalt and iron sorbed on Dowex 1, and by Klier and Ralek (2) on nickel ions sorbed on synthetic zeolites. The samples were generally placed in a vacuum cell provided with a quartz window (2) and its reflectance R, relative to that of a MgO standard placed in a matching cell, was recorded. As the absolute reflectance (ratio of intensities of reflected and incident light) of MgO is, within the spectral range investigated, close to unity, the values of R represent, with good accuracy, the absolute reflectance of the specimen. The absorptivity was evaluated from the relation given by Kortüm et al.(3).

$$f(R_\infty) = \frac{(1-R_\infty)^2}{2R_\infty} = \frac{k}{s}$$

where

R_∞ is the relative reflectance of a layer of infinite thickness with respect to the absolutely white standard (R_∞st = 1); infinite thickness was approximated by a layer 3 mm thick

k, the absorption coefficient and

s, the scattering.

Similar techniques have more recently been used for study of the coordination of Co^{2+} and Ni^{2+} in zeolites.(5)

Coleman (6) used another technique to study the chlorocomplexes of cobalt. He prepared flattened discs from individual resin spheres, 0.2 - 0.6 mm thick, placed them in a hole in platinum foil and then taped the foil into a standard 1 cm cell. The resin spheres, previously equilibrated with $CoCl_2$, were exposed to HCl fumes. From his results it is not clear whether he obtained the spectrum of the sorbed species or that of the species adsorbed only on the surface.

Ryan (7,8) measured the absorption spectra of complex species of the uranyl ion sorbed on an anion exchanger using 0.5 cm or 1.0 cm cells containing the resin settled in the equilibrium solution. A fine-tipped pipette was used to remove excess liquid until the resin, so packed, could not move in the thermal gradients produced by the light beam. The same resin free of metal was used as the blank.

Nortia and Laitinen (9-12) measured the spectra of resins, air dried or moist, in 2 mm cells. In another approach resins dried at 100°C, mixed with nujol to a paste, were mounted on filter paper. Although this last procedure is generally the recommended one for solid compounds, it is not very useful for organic ion exchangers which are very difficult to grind to a homogeneous powder.

Waki et al. (13) packed resin beads with some of the equilibrium solution in a 0.5 or 1.0 mm space made by inserting a quartz spacer into a 1.0 cm quartz cell. A net absorbance $A_{B,C}$ was calculated with the following relation:

$$A_{B,C} = A - A_{soln} - A_B$$

where

A is the overall measured absorbance of the sample;
A_{soln}, the absorbance of the equilibrium solution; and
A_B, the absorbance of the resin itself.
A and A_B were determined in two separate measurements.

In the work done in our laboratory (14), a somewhat different technique was adapted and found suitable for the evaluation of quantitative parameters inaccessible in the earlier studies. The technique consisted of preparing a suspension of air dried resin beads (50-100 mesh) in nujol. The suspension, introduced into 1 mm quartz cells, was then centrifuged for 5 minutes to obtain a homogeneous compacted sample for spectral study. The same exchanger, free of metal ion and prepared in the same way, was used as the reference.

Molar extinction coefficients (ε_{mol}) were evaluated with the standard equation, $\log I_o/I = \varepsilon \ell c$. The value of c for the species sorbed was obtained by transforming the loading of the resin from milliequivalents per gram to moles per liter of exchanger. The volume of 1 g of exchanger was found to be the same in air as in nujol.

The oscillator strength was computed: $f = 4.6 \times 10^{-9} \varepsilon_{max} \Delta\nu_{1/2}$ where ε_{max} is the molar extinction coefficient and $\Delta\nu_{1/2}$ is the half intensity band width, i.e. the width at $1/2\,\varepsilon_{max}$. Other terms evaluated from the visible spectra were the 10 Dq - crystal field splitting- and B-nephelauxetic-parameters (15).

B. Infrared Spectra

Most of the IR measurements on zeolite samples have been made in the standard way; KBr discs (16,17) or thin wafers 1" in diameter (18,19,20), were prepared from powdered zeolites, the weight of the sample being 0.05 g.

The quality of the IR spectra of organic exchangers made in this manner depends strongly on the grinding state of the sample. Good results were obtained with nujol mulls and KBr pellets of very finely ground resin.

Zündel measured the IR spectra of thin ion-exchange membranes (21,21a) in a special cell designed to control temperature and humidity.

C. Raman Spectra

In recent years Raman Spectroscopy (22-25) has been used to good advantage for the study of zeolites and organic ion exchangers. With this method, the scattering by water is small; measurements can also be made in spectral regions where water absorption bands interfere when using the IR method. The Raman method is very suitable for measurements at small frequencies where most metal-ligand vibrations occur. Samples can be very small for Raman measurements when gas lasers are used for excitation.

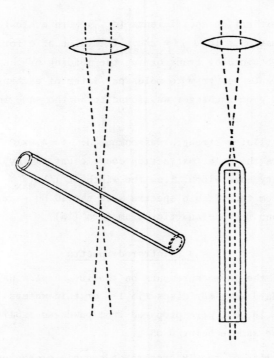

Fig. 1. Sample arrangements for Raman spectrum excitation: axial (left) and transverse excitation (right) (23).

Dobud et al. (22) have studied the halide species of gold and indium sorbed on anion exchangers. In these studies, the resin was equilibrated with the appropriate solution, washed, vacuum dried, and then ground. The dried material was packed into a standard conical sample holder for coaxial viewing in the spectrometer.

Angell (25) prepared pellets of zeolite and examined them at 90° illumination. Boyd (23) performed his measurements on very small samples to accommodate the small diameter and non-divergent character of laser light beams (Fig. 1). He described two different arrangements of the samples for excitation. Difficulties that were encountered in the examination of cation exchangers were apparently due to excessive scattering effects that arose from

SPECTROSCOPIC STUDIES OF ION EXCHANGERS

heterogeneities within the bead. Another operational problem with cation exchangers is due to fluorescence, especially when excitation is conducted with a blue laser.

D. Other Methods Used

NMR, ESR and magnetic susceptibility measurements are particularly suitable for the evaluation of fundamental properties of ion exchangers and zeolites. The standard sample holders may be used and the measurements may be performed on dry or hydrated material. As will be seen, important additional quantitative data are easily available from these measurements.

Mössbauer effects have been studied on compressed wafers of zeolites (26,27) as well.

III. SPECTROSCOPIC STUDIES ON ZEOLITES

A. General Properties and Uses

Zeolites are hydrated crystalline aluminosilicates which can be reversibly dehydrated without essential changes in structure and may act as cation exchangers.

Zeolite composition corresponds to the following general formula:

$$(M^I\, M^{II}_{1/2})_x [xAlO_2 \cdot ySiO_2] \cdot zH_2O \quad \text{where}$$

M^I = Li, Na, K. etc. and

M^{II} = Mg, Ca, Sr, Ba, etc. normally $(x \leq y)$.

On considering their structure they may be described as a cross-linked polymeric macromolecule in which almost all the oxygens bridge the Si and/or Al atoms. At the crystal faces one oxygen is not bridged but is linked to a proton (forming a hydroxyl group). Due to the presence of trivalent aluminum, there is a negative charge of one unit per atom. These charges require the interstitial arrangement of alkaline or alkaline earth cations for formation of the neutral crystal. These cations are rather mobile and may be exchanged by other ions; i.e. they are responsible for the cation-exchange properties. Water molecules fill the cavities in the framework. The arrangement of the $(Si,Al)O_4$ tetrahedra are different and the rings or cages thus formed control the size of the ion exchanged or the molecule sorbed; hence, the name given to these zeolites: molecular sieves.

The determination of the structure of the zeolites presents many difficulties for the following reasons. The cations and water molecules are rather mobile and often not fixed to their sites in the framework. Thus, exchange processes may cause their transfer to other sites. The usual procedure for the determination of structure by isomorphic replacement can, therefore, not be used. Through X-ray and electron diffraction studies the framework of many zeolites (natural and artificial) has been determined and a classification of zeolites according to structure type has been proposed by Smith (28), Fischer and Meier (29) and Meier (30). This classification is based on a similarity of framework topology and common elements of secondary building units which are comprised of tetrahedral rings, double rings and larger symmetrical polyhedral units, such as the 18-tetrahedra "cancrinite" unit or the 24-tetrahedra truncated octahedron "sodalite" unit (30). Because of the large unit cells and the different ways that the tetrahedra are linked, the X-ray technique is not always best for determining the position of the atoms in the zeolites. Furthermore, crystals sufficiently large for successful application of X-ray techniques are not always obtainable with artificial zeolites. It is at this point that the other spectral techniques have been introduced to facilitate these studies. They have contributed very much to the further elucidation of zeolite structures and reaction mechanisms in the zeolite.

SPECTROSCOPIC STUDIES OF ION EXCHANGERS

B. The Structure of the Framework

Most of the studies performed to elucidate the framework structure have been based on infrared measurements (31-35). Zdanov et al. (34) measured the IR spectra of a series of Na-faujasites in which the Si:Al ratios vary from 1.11 to 2.55 (Fig. 2). It is

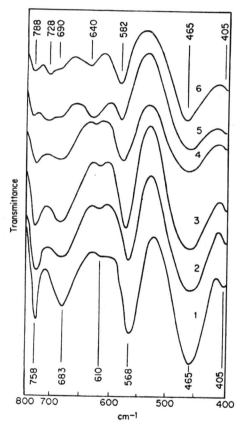

Fig. 2. Infrared vibrational spectra for crystalline skeletons of synthetic faujasites of various compositions after vacuum treatment at 400°C for four hours. Ratios Al/Si: curve 1 - 1.11; 2 - 1.28; 3 - 1.50; 4 - 1.73; 5 - 2.31; 6 - 2.55 (32).

Table I. Infrared Spectral Data for Synthetic Zeolites[a]

cm^{-1}

Zeolite	$\frac{SiO_2}{Al_2O_3}$	Asym. stretch				Sym. stretch		
A	1.88	1090vwsh	1050vwsh	995s				660vw
$Ca^{ex}A$	1.9	1130vwsh	1055vwsh			742vwsh	705vwsh	665vw
N-A	3.58	1131vwsh		1030s		750vwsh		675vw
N-A	6.01	1151vwsh		1044s		750vwsh		698vw
X	2.40		1060msh	971s		746m	690wsh	668m
Y	3.42	1135msh		985s		760m		686m
Y	4.87	1130msh		1005s		784m		714m 635vw
$La^{ex}Y$	5.0	1135msh		1006s		790m		705m
Y	5.63	1130msh		1017s		789m		718m 645vw
B(P)	2.8	1105mwsh		995-1000s		772mwsh	738mw	670mw
Hydroxy sodalite(HS)	2.0	1096vwsh		986s		729m	701mw	660m
Ω	7.7	1130wsh		1024s		805mw	722mw	
ZK-5	6.0	1158wsh		1048s	890vwb		730mw	
R	3.25	1136mwsh		1007s		738w	678w	
G	5.44	1138mwsh		1027s		720w	696wsh	
D	4.62	1184mwsh		1018s		755wsh	711w	
S	2.5	1140wsh		1020s		770vwsh	722mw	690vwsh
T	7.0	1156wsh	1059s	1010s		771w	718w	
Hydroxy cancrinite(HC)	2.0	1095mw	1035msh	1000s	965msh	755w	680m	
L	6.0	1160wsh	1080s	1015s		767mw	721mw	642vwsh
C	4.0	1162vwsh	1012s	952s		740m	686wb	
Zeolon	9.95	1216w	1180vwsh	1046s		795⎫wb 772⎭	715⎫wb 690⎭	
W	3.6	1128msh		1006s		786⎫mwb 756⎭	691mwb	

SPECTROSCOPIC STUDIES OF ION EXCHANGERS

	Dbl. rings		T-O bend			Pore opening?	
	550ms		464m			378ms	260vwb?
	542ms		460m			376m	
	572ms		474m			385m	
	581ms		475m			393m	
	560m		458ms		406w	365m	250vwb
	564m	508vwsh	460ms			372m	
	572m	500wsh	455ms			380m	260vwb?
	565m	500wsh	450ms			382m	
	575m	504mwsh	456ms			383m	315vwsh
600m				435ms		380mwsh	
			461ms	432ms			282vwb?
610mw			451ms			372m	
	572m		445m		408wsh		
625m		508mw	452m	426m		370vwsh	
632m		515m	460m	408m		378vwsh	
631m		513m	459m	415m		376vwsh	
623 }mb 595sh		518mb	448m	424ms		370vwsh	
623mw	575w		467ms	433ms	410vwsh	366wsh	
624m	567m	498mw	458ms	429ms	390mw	353wb	
606m	580wsh		474ms	435wsh		375vwsh	
515vw			442ms	410msh			
621w	571 }w 555		448ms			370vwsh	
637mw	590wb	512vwsh	483vwsh	432ms		375vwsh	

[a] S = strong; ms = medium strong; m = medium; mw = medium weak; w = weak; vw = very weak; sh = shoulder; b = broad.

known from X-ray data that these changes in composition have little effect on the lattice constant of these zeolites; however, they cause a twofold decrease in exchangeable cations per unit cell. These changes also cause some regular shifts in the IR spectrum. As may be seen from Fig. 2, the bands located at 568, 610 and 758 cm^{-1} are particularly sensitive to the changes in composition. In the same study, effects of the charge and radius of the compensating cation were studied. The IR spectra of zeolites are different when the cation is a monovalent alkaline ion or a divalent alkaline earth cation. The band sensitive to these changes is at 763 cm^{-1}.

The most detailed study of the framework structure of zeolites and their classification into groups has been made by Flanigen et al. (36). This study was made in the IR spectral region of 1300-2000 cm^{-1} since it contains the fundamental vibrations of the framework (Si,Al)O$_4$ tetrahedra. A variety of artificial zeolites was examined and Table I presents the main features of the spectra obtained. A correlation has been made between the spectra and the zeolite structure. Each zeolite type is characterized by a typical infrared spectrum. In the spectral region studied, two classes of

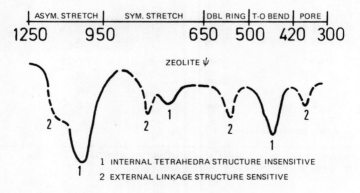

Fig. 3. Infrared assignments illustrated with the spectrum of zeolite Y, Si/Al of 2.5 (36).

vibration were found. The first class is due to the internal vibrations of the TO_4 tetrahedron (T is Si or Al), the primary building unit of the zeolite. These vibrations are not sensitive to changes in framework structure. The second class of vibrations is related to the external linkage between tetrahedra and thus is sensitive to the structure of the framework. Flanigen et al. (36) do not specifically assign vibrations to SiO_4 or AlO_4 but rather to an average TO_4 group and T-O bond where the vibrational frequencies represent the average Si,Al composition and bond characteristics of the central T cation. These assignments are summarized in Table II and Fig. 3.

Table II
Zeolite Infrared Assignments, cm^{-1} (36)

Internal Tetrahedra	
Asym. stretch	1250-950
Sym. stretch	720-650
T-O bend	420-500
External Linkages	
Double ring	500-650
Pore opening	300-420
Sym. stretch	750-820
Asym. stretch	1050-1150 sh

As may be seen from this figure and the table, the internal vibrations are assigned to the stretching modes situated between 950 and 1250 (asym) and between 820 and 650 cm^{-1} (sym). The symmetric stretch has been classified into an internal linkage symmetric stretch in the lower region of 650-720 cm^{-1} and an external linkage stretch which is sensitive to structure in the higher region of 750-820 cm^{-1}. All the stretching modes are sensitive to the Si/Al ratio and shift to a lower frequency with a decrease of Si/Al. The T-O bending mode is not especially sensitive to the Si/Al ratio. These assignments are in good agreement with those made previously for pure silicates (37).

The assignments of Flanigen et al., (36) as was pointed out above do not take into consideration the internal tetrahedral vibrations of AlO_4, considered by earlier workers (35,38). Their justification for this omission in the analysis of their IR data is the fact that the relative concentration of Si and Al alters the frequency of the band but not the number of them. This has been confirmed by X-ray studies in which the equivalency of Si and Al was found and the T-O bond lengths reflected the average composition.

The external linkage frequencies are those which are sensitive to the structure of the building units in the framework. They were found in the regions of the spectrum at 500-600 cm^{-1} and 300-420 cm^{-1}. A band of medium intensity in the 550-630 cm^{-1} region, related to the presence of double (D) ring polyhedra, has been found in all the zeolites with D-4 and D-6 rings (X,Y,B,ZK-5) and in the chabazite group phases. Zeolites without D-R or larger symmetrical polyhedra show only a weak absorption in this region.

A somewhat less certain assignment is made for the 300-420 cm^{-1} region. The band appearing here is tentatively assigned to a breathing motion of the isolated rings forming the pore openings in zeolites. For the larger symmetrical polyhedra (18-24 tetrahedra) no vibrations are expected above 300 cm^{-1}. A careful study (36) (in CsI wafers) of the 200-300 cm^{-1} region showed a broad band tentatively connected to these building units.

A shift in frequency of some bands is obtained with changes in Si/Al ratios. As may be seen from Fig. 4, a linear decrease in frequency is obtained with an increase in the Al fraction in the framework. This frequency decrease results despite the very similar atomic weigths of Si and Al; however, the bond length of Al-O is longer and its electronegativity smaller, thus causing a decrease in the force constant.

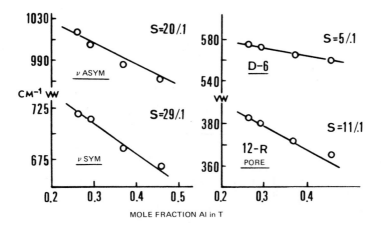

Fig. 4. Frequency vs. atom fraction of Al in the framework for zeolites X and Y for several infrared bands (36).

On the basis of Flanigen's data (36), Lahodny-Sarc (16) has interpreted the IR spectra of aluminum-deficient zeolites obtained by treatment with acid or EDTA. His results are in good agreement with Flanigen's, and the increase in the Si/Al ratio upon aluminum removal is believed to produce a distortion in the framework and changes in the symmetry of the ring cages. Another structural study of the zeolite framework based on IR measurements was recently made by Kubasov et al. (18). In their study IR spectra of decationated zeolites were measured following heat treatment at different temperatures (Fig. 5). As may be seen from this figure, the major change occurs at about 580 cm^{-1}, where the band becomes broader until it disappears as a result of longer heat treatment. From the X-ray structure, it may be seen that upon heating, the samples become less crystalline, or even amorphous. Changes in the 580 cm^{-1} IR band were also linked by Flanigen et al. to a loss of structure of the zeolite.

These studies (36,16,17) have provided valuable data for the classification of zeolites. The extension of these studies promises

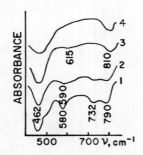

Fig. 5. Infrared spectra of the "decationated" HY-1 zeolite after preliminary heat treatment in air at different temperatures: 1 - 380°C, 4h; 2 - 550°C, 2h; 3 - 550°C, 8h; 4 - amorphous "Houdry" aluminosilicate (zeolite/KBr = 1/100) (17).

further refinement of the information already obtained. The spectra are rather simple and more specific assignments, with additional study by this approach, seems likely. Finally, with the development of new lasers, it will be possible to extend the utility of Raman spectral techniques. The Raman technique has already been applied by McNicol et al. (24) to monitor zeolite synthesis. The Raman band in the solid phase has been attributed to the binding of a $(CH_3)_4N^+$ ion to the negatively charged Al framework sites of the amorphous gel.

C. Water and Hydroxyl Groups in Zeolites

Water molecules and cations are found in the cavities of the framework and, as has already been pointed out, hydroxyl ions are formed on the surfaces. The type of cation linkage is closely connected to the H_2O and OH^- content of the zeolite. In general, most authors compare the cations and the water in the cavities of a zeolite or ion exchanger to a concentrated aqueous salt solution (39-43). Support for this assumption is found, mostly for open-structured zeolites, in data obtained from IR spectra, conductivity, self diffusion, NMR and other measurements. In less open structures (44), this assumption is no longer supported by the much lower diffusion coefficients measured (45). Upon dehydration of the framework, the zeolites act as a molecular sieve and various molecules are adsorbed.

It is not the purpose of this review to present a complete survey of this subject. However, some of the studies in which NMR and IR spectra techniques were employed have contributed to a better general understanding of this topic and are included in our discussion.

For a fully hydrated Na^+ form of a faujasite (43), the stretching and deformation frequencies were found to be the same as those for liquid water. In the series of zeolites containing alkali ions, the effect of dehydration was studied by IR spectra. In most cases, no significant changes were found, e.g. a 10-20 cm^{-1} increase in the main frequencies. With the less open-structured zeolites in the divalent or trivalent cation form, however, dehydration generally caused rather strong distortions in the framework and, consequently, changes in the IR spectra. In the more open-structured zeolites exchanged with polyvalent cations, no significant changes in the IR spectra were found as there is almost no interaction between the cation and the framework (see Table I).

The surface properties of zeolites have been discussed in a series of papers by Ward (46,18,19). IR spectra provide useful information as to how these surfaces are modified by various treatments. Three types of hydroxyl groups were detected by Bertsch and Habgood (47). The frequencies of the bands detected were found to increase with an increasing electrostatic field (electrostatic and ionization potentials were calculated for the cation form of the zeolite). In most of the zeolites, no hydroxyl ions could be detected after dehydration, apart from the group IIA zeolites (48).

From these studies, a coherent picture of the surface was obtained. In the absence of cation deficiency in monovalent zeolites no structural hydroxyl groups are observable. When hydroxyl groups are detected, they result from cation deficiency caused by partial hydrolysis or by small amounts of siliceous impurities. The surface of all other zeolites contains structured OH^- groups with absorption bands at 3690 cm^{-1}, 3650 cm^{-1} and 3540 cm^{-1}. The concentration of these groups is much greater than observed in Group IA zeolites of the cation-deficient type. It is therefore unlikely that the presence of OH^- groups in the IIA zeolites results from cation deficiencies. Their appearance is presumed to arise in the following way: A simple divalent cation apparently cannot satisfy the charge distribution requirements of the zeolite lattice in the absence of water. During hydration, the multivalent cation is localized and its associated electrostatic field may induce dissociation of coordinated water molecules to produce MOH^+ and H^+ species. The proton then reacts with a lattice oxygen at a second exchange site to produce the type of OH^- that is usually found in hydrogen zeolites; i.e. with absorption frequencies of 3650 and 3540 cm^{-1}

the third band at 3690 cm^{-1} was originally assigned to $AlOH^{2+}$ groups and, more recently, has been assigned to physically adsorbed water [48].

Numerous IR studies have recently been made to attempt correlations between surface structure of the zeolites and their catalytic properties. By the NMR technique, Resing and Thompson (45) have measured the state of water condensed in the pores of zeolites at temperatures ranging from 200° to 500°K. The time between molecular jumps or correlation time, τ, was employed to measure the fluidity of water adsorbed to saturation. For a bulk liquid the correlation time, τ, was related to the viscosity η by an equivalent of the

Debye equation. Such analysis of the data shows that the intracrystalline fluid is about 30 times as viscous as bulk water at room temperature and is as dense as liquid water.

D. Ion Exchange and Catalytic Properties

The cation-exchange properties of crystalline zeolites are essentially different from those reported for the organic ion exchangers. Zeolites do not swell, there are very significant differences in selectivity coefficients, and specific sieve effects are encountered because of their well defined, narrow channels. Their use is limited by their instability towards acids and bases. Many equilibrium and kinetic studies of the exchange process in these materials are not pertinent to the objective of this chapter and are not considered here. However, many spectral studies that have been carried out are germane to this discussion. They have enabled the identification of the cationic species sorbed. The techniques that have been employed for this purpose are UV, visible, and NIR as well as ESR and Mössbauer spectroscopy. Through such knowledge, insight with respect to the structure of the zeolite cavity can be obtained. These measurements enable characterization of the cationic species from the point of view of coordination, ligand field symmetry, and type of bonding. The catalytic properties of the zeolites depend on the cation sorbed and the amount of water present. The transition metals which exhibit specific catalytic properties and are very susceptible to spectral study as well are thus most suitable candidates for such analysis. For appropriate development of this section of the article, spectral properties of the first row transition element forms of the zeolite that have been measured are presented.

1. Manganese

Faber and Rogers (50) were among the first to study the ESR spectra of transition metal ions sorbed on ion exchangers and zeolites. The spectra of the manganese species (probably the hexahydrate) showed a very poorly resolved hyperfine structure with

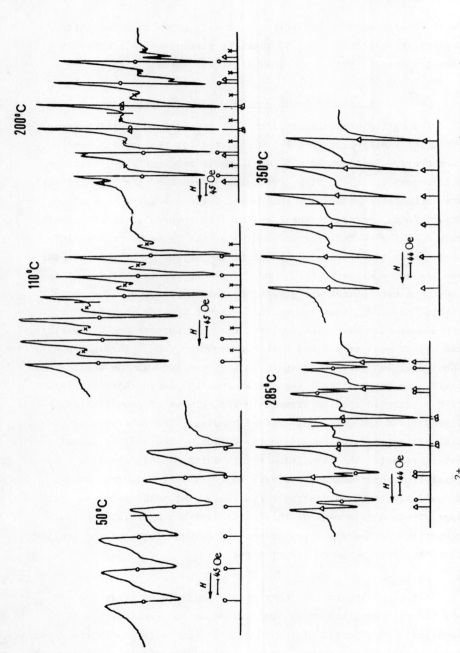

Fig. 6. ESR spectra of Mn^{2+} in zeolite NaY/Mn after vacuum heat treatment at various temperatures. The line in the central part of the spectra indicates DPPH line position (52).

$g = 2.00$ and a hyperfine structure interval of $A' = 96 \pm 3$ gauss. These values for g and A were resolved for divalent manganese in both ionic crystals and aqueous solutions, and it may therefore be concluded that the binding in the zeolite is also essentially ionic.

Barry and Lay (51) have studied the ESR spectra of Mn sorbed on a hydrated zeolite prior to and after heating to 200°, 300°, 400° and 600°C. From the changes that were observed in the g and A values, an increase, with zeolite dehydration, in the covalent character of the binding as well as changes in symmetry were inferred. The bound species were identified as $MnOH^+$ and MnO, formed by hydrolysis during dehydration. These changes in g and A were also used to deduce the sites occupied by the cation.

A similar ESR study of manganese in zeolites has been performed by Dzhashiashvili et al. (52). A very good resolution of the hyperfine structure was obtained at all degrees of dehydration (Fig. 6). From these spectra the g and A values were calculated. The decrease in A from 95 to 87 gauss through the dehydration of zeolite A was also attributed to an increase in covalency with the loss of water molecules. The manganese ion is thus assumed to bind directly to the zeolite lattice. The decrease in hyperfine splitting may also be caused by a change in the symmetry of the field from octahedral to nearly tetrahedral. In other types of zeolites like zeolites Y and X, a spectrum with a much higher hyperfine structure interval was found upon dehydration (99 gauss). This result was explained by the fact that the manganese ion is in an electric field of very high symmetry. The location of the cations in dehydrated zeolite X and Y was assigned to 6-membered oxygen rings having a small cavity or sodalite cell. The manganese ion may thus be at the center of a hexagonal prism inside the sodalite cavity.

2. Iron

Iron in zeolites has been studied principally through the Mössbauer effect (26,27,53). Figure 7 depicts representative spectra obtained for a zeolite under various experimental condi-

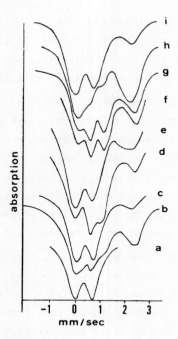

Fig. 7. Iron in Na-Y a) as prepared by ion exchange; b) after outgassing at 360°C; c) after exposure to atmosphere; d) reoutgassed at 360°C; e) after sorption of methyl alcohol; f) reoutgassed at 360°C; g) after contact with oxygen; h) after partial rehydration; i) after exposure to atmosphere (53).

tions (53). It is clear from this study that valency changes occur during dehydration of iron zeolites, i.e. a spectrum characteristic of ferrous ion is obtained. Some of the spectra are much more complicated and indicate coexistence of the ferric and ferrous states. Such valency changes may be closely linked to their catalytic properties.

Delgass et al. (26,27) have studied the Mössbauer spectra of Fe^{2+} adsorbed on Dowex 50 and Linde 4 zeolites. The water-saturated solids were evacuated at room temperature or cooled to liquid nitrogen temperature prior to their examination. The spectra thus obtained were characteristic of high spin Fe^{2+}. The effect of hydration was explained in terms of solvation and mobility of the iron ions in a diffuse double layer in the wet resin. The qualitative picture obtained (26) was analogous to that derived from ESR spectra for the manganese ion (52). In the fully hydrated zeolite, some of the iron was believed to be solvated and mobile in the intracrystalline fluid. When half of the water was removed, localization of the Fe^{2+} ions was considered to be complete with the formation of strong lattice bonds. In all the dehydration experiments, including even those in which the zeolite sample was heated under vacuum at 400°C for 14 hours, rehydration resulted in complete solvation. According to the authors, this is evidence for a true chemical exchange reaction that can be used as a test for reversible exchange. The objective of the second paper of Delgass et al. (27), the determination of the local environment of the iron ion, was accomplished by measuring the Mössbauer spectra of the iron in zeolites when gases of various sizes were sorbed (O_2, N_2, CO, NO, C_2H_4, CS_2, t-butyl-alcohol, pyridine, piperidine, and NH_3). The gases used can be classified into three groups according to their adsorption and the spectra obtained. The inner and outer peaks of the spectrum of the dehydrated Fe^{2+}-Y zeolite that was obtained showed iron to exist in at least two different environments. The outer peaks are similar to those found for octahedral configuration of oxides. (Such coordination is possible in zeolite Y only in the center of the hexagonal window and/or of the hexagonal prism.) The first possibility, ruled out by the low values of the physical adsorption of krypton on Fe^{2+}-Y, leads to the assignment of the outer peaks to iron ions in the hexagonal prisms. The parameters of the inner peaks are very sensitive to the adsorbed gases, showing that these iron ions are accessible to gases of different sizes. The inner peaks, though much more difficult to interpret, are apparently not due to iron ions outside of the hexagonal prisms. A diagram of a hexagonal window site is presented in Figure 8. A small cation like Fe^{2+} can well compensate for the charge due to Al. It has been determined (46) from IR spectra that, with dehydration, the

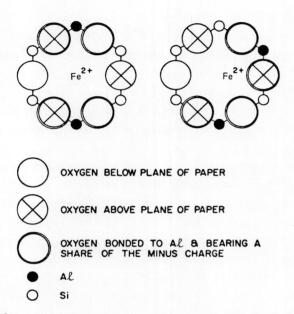

Fig. 8. Fe^{2+} in the hexagonal window site of Y zeolite (27).

species $M^{2+}(OH)^-$ is formed through dissociation of a water molecule. By such consideration of the hexagonal window, one can see that if an Al-O-Si bridge is broken to form AlHO-Si (see page 100), then $Fe^{2+}(OH)^-$ can compensate for the charge from the remaining Al. Thus the Fe^{2+} can move from sixfold coordination to tetrahedral (Fig. 9); shifts of the inner peaks are consistent with this estimate and also are in agreement with the observed property of other minerals tetrahedrally coordinated. It is through such data that valuable information as to the chemical nature of the atoms on the zeolite surface has been obtained.

3. Cobalt

The coordination of cobalt ions in zeolite X and Y has been studied by their reflectance spectra (4) and magnetic susceptibilities. (54) The preferential change in symmetry from octahedral to tetrahedral coordination is accompanied by very significant spectral

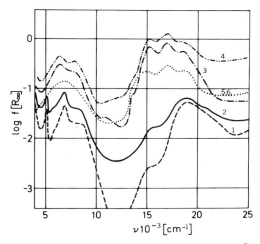

Fig. 9. Compensation of an AlO_4^- tetrahedron by $Fe^{2+}(OH)^-$ (27).

changes to make this technique a most effective one for this kind of study. Figure 10 shows the spectra of cobalt ions in a hydrated, a partially hydrated and a fully dehydrated sample. The spectrum of the hydrated zeolite is identical to that of an aqueous solution of cobalt $[Co(H_2O)_6^{2+}]$; i.e. the sample has the characteristic pink color. Upon partial dehydration, the color of the sample changes to blue and the bands characteristic of tetrahedrally coordinated cobalt are found. There is no spectral indication of water molecules. In the hydrated zeolite it can be assumed that the cobalt ions are not fixed in any particular position in the zeolite but "float" in the intracrystalline fluid (water solution), while in the partially dehydrated zeolite (350°C) the tetrahedral environment of the cobalt ion is apparently due to three oxygen atoms from the framework and one water molecule fragment, OH^- or O^{2-}. After further heating of the zeolite to 600°C, the tetrahedral bands are no longer well resolved; it is believed that the Co^{2+} ions are contained by the hexagonal prism, inside the sodalite cavity, the octahedral environment of the oxygen atoms of the zeolite framework redetermining the spectral properties. The magnetic suscept-

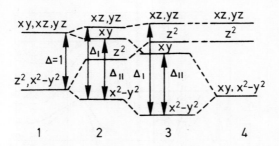

Fig. 10. Reflectance spectra of Co zeolite X: 1 - original; 2 - 350°C + H₂O(25°C); 3 - 350°C; 4 - 350°C + O₂(300°C); 5 - 600°C; 6 - 600°C + O₂(25°C). (4)

ibility measurements of Egerton et al. (54), as well as the X-ray study of Gallerot and Imelic (55), are in very good agreement with these conclusions.

The coordination of cobalt in zeolite type A samples was studied by Klier (56,57). The cobalt ions sorbed on a hydrated zeolite A sample yield a spectrum identical with that of the zeolite X or Y, i.e. the $Co(H_2O)_6^{2+}$ species is sorbed. Dehydration at 350°C, however, produces a species different from the tetrahedrally coordinated Co^{2+} in the partially dehydrated zeolite X and Y (4,54,55). A qualitative assessment of the spectrum indicates that the bands are characteristic of trigonal symmetry D_{3h}; the basis for this estimate is the spectral property of $d^1(d^9)$ and $d^2(d^8)$ electron systems of this symmetry.

Co(III) (d^7) System

	$d_{x^2-y^2}, d_{xy}$	(e')
	d_{zx}, d_{yz}	(e'')
	d_{z^2}	(a_1')

free ion D_{3h} field

The highest energy band of 24,000 cm^{-1}, $a_1' \rightarrow e'$, is assigned to the transition from the d_{z^2} orbital, with maximum electron density on the axis perpendicular to the plane of oxygens, to the in-plane, partially occupied d_{xy} and $d_{x^2-y^2}$ orbitals. The intense bands at 15,000-20,000 cm^{-1} and at 7000 cm^{-1} are assigned to the remaining two transitions $e'' \rightarrow e'$ and $a_1' \rightarrow e''$ without specification. As may be seen, all transitions are spin allowed, one electron transitions.

It is assumed that the Co^{2+} ions are situated in S-II-trigonal sites. Cobalt ions in these sites are capable of adsorbing the following gases: H_2, O_2, ethylene oxide, nitrous oxide, cyclopropane, water and ammonia. Hydrogen and oxygen do not effect the spectra and it may be assumed that the cobalt ions sorbed are resistant to reduction and oxidation, the opposite of what happens to oxides and hydroxides which are easily oxidized. If either N_2O, cyclopropane, water or ammonia is sorbed, changes in the spectrum occur. The band at 24,000 cm^{-1} is eliminated or shifted to lower energies and the band at 7000 cm^{-1} becomes somewhat split. The color changes from pale blue to violet. After desorption of all adsorbates except ammonia, the original dehydrated spectrum of Co is obtained.

The shift of the 24,000 cm^{-1} band is attributed to the repulsive interaction between the electron pair of the ligand and the out of plane d_{z^2} orbital of the Co^{2+} ion, the magnitude of the shift being determined by the strength of the ligand. Water and ammonia produce tetrahedral complexes and are the weakest ligands. The N_2O and cyclopropane are stronger ligands and apparently the symmetry of the complex formed is C_{3v}, the molecule of adsorbed gas being on the z axis perpendicular to the plane of the oxygens. The spectrochemical series for these ligands is: $N_2O < C_3H_6 < H_2O < NH_3$. The N_2O molecule may be desorbed by pumping at 50°C; however, if the desorption is effected at 150°C, the nitrous oxide decomposes into a mixture of $2N_2 + O_2$. The cobalt ions which remain unchanged apparently catalyze this reaction. This reaction most likely has as a first product a N_2 molecule and an oxygen atom and should have an important role in selective oxidation catalysis.

4. Nickel

Klier et al. (2,58,59) have reported spectral studies of the nickel zeolites as well. Results analogous to those described for cobalt (56,57) were obtained. Guilleux et al. (60) have followed the dehydration of the NiA and NiX zeolites by infrared measurements. The spectral bands obtained for NiA agree with the results obtained by Klier and Ralek. For NiX, evidence for formation of $NiOH^+$ was obtained to duplicate this observation for other divalent ions and confirm the dissociation of water molecules as follows:

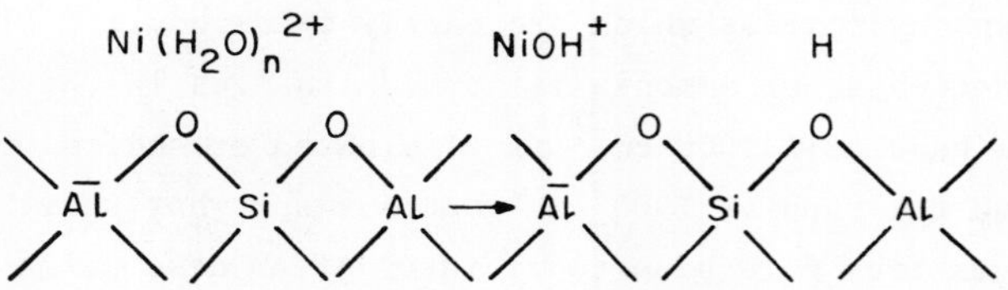

$$Ni(H_2O)_n^{2+} \quad\quad NiOH^+ \quad\quad H$$

[Diagram of zeolite framework with Al and Si tetrahedra bridged by O atoms, showing migration arrow from left Al site to right Al site]

Garbowski et al. (5) have a different interpretation of results obtained in their study of nickel ions sorbed on zeolites X, Y and A. The hydrated nickel, $Ni[H_2O]_6^{2+}$, is found in the open cavities; upon heating the sample in vacuum, the cations migrate into the hexagonal prism where they coordinate with the oxygen atoms from the framework. The visible spectra were interpreted by resort to crystal field theory and in all cases octahedral configurations were assumed.

5. Copper

Most of the studies of Cu^{2+} zeolites have been made through ESR measurements (61-67) and only one or two papers deal with their visible (61) and infrared spectra (65). A critical ESR study and some visible spectra studies have brought Mikheikin et al. (61) to the following conclusions: Copper ions are sorbed on a hydrated

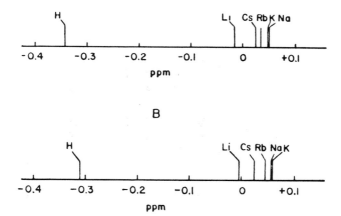

Fig. 11. Scheme of energy levels for a 3d hole in various crystal fields: 1 - octahedral; 2 - square pyramidal; 3 - square planar; 4 - triangular planar (61).

zeolite Y as $Cu(H_2O)_6^{2+}$ and their ESR and visible spectra are identical to that of the ion in solution. These hydrated ions occupy the large cavities and have a high mobility. On heating the zeolite at 200-300°C most of the water is removed and the mobility of the copper complex is greatly reduced, to suggest a strong binding to the oxygens of the zeolite framework. The configuration of the copper species is assumed to be square pyramidal or square planar and the g values obtained are consistent with this assumption. On heating the zeolite samples at more than 400°C a strong decrease in the intensity of the ESR spectra is observed. This result may be attributed to a corresponding change in symmetry, which is possible if copper occupies trigonally-distorted 6-member rings. From the energy level scheme given in Fig. 11, it may be observed that in this case the ground state of the Cu^{2+} ion is orbitally degenerate. A slight removal of this degeneracy (Jahn-Teller effect) leads to a very large broadening of the ESR lines.

In a very detailed ESR spectral study (63) of copper ions in Y zeolites in the presence and absence of water and other adsorbates,

Table III

Magnetic and Bonding Parameters (66)

	$g_{\parallel}$	$g_{\perp}$	$A_{\parallel} \times 10^4$ cm^{-1}	$A_{\perp} \times 10^4$ cm^{-1}	α^2	β_1^2
Cu(acac)$_2$ in C$_6$H$_5$CH$_3$	2.246	2.062	197	15	0.82	0.73
Cu(acac)$_2$ on SiO$_2$	2.253	2.055	195	28	0.86	0.72
Cu(acac)$_2$ on zeolite	2.264	2.044	190	31	0.85	0.75
Cu(acac)$_2$ on SiO$_2$-NH$_3$	2.240	2.029	193		0.83	0.71
Cu(acac)$_2$ on zeolite-NH$_3$	2.268	2.034	183		0.83	0.79
Cu(acac)$_2$ on SiO$_2$-C$_6$H$_5$NH$_2$	2.301	2.031	160		0.80	0.91
Cu(acac)$_2$ on zeolite-C$_6$H$_5$NH$_2$	2.300	2.054	165	23	0.82	0.89

the following deductions were made: In the hydrated zeolite, $Cu(H_2O)_6^{2+}$ is found in the large cavity. The data obtained can also be ascribed to copper clusters. Together with this type of copper species, a second copper species is apparently localized on the walls of the zeolite. These estimates are based on the existence of a symmetrical signal with $g = 2.17$ and an asymmetric signal with $g_\perp = 2.10$ and $g_\parallel = 2.38$ in the ESR spectrum. With a decrease in temperature, the number of localized Cu^{2+} ions is increased while the concentration of free tumbling ions is decreased; the symmetric peak even disappears. In the dehydrated zeolite, copper clusters are assumed. Support for this assumption is provided by the fact that an increase in the copper concentration increases the ESR signal. The existence of two different signals in the dehydrated zeolites is attributed to copper ions localized in two different sites; however, no attempt is made to define these sites.

An ESR study of bis(acetylacetonato) copper adsorbed on zeolite X has been made by Yamada (66,67). From his measurements g and A values were determined, and from them the bonding parameters α^2 and β_1^2. Table III presents these parameters for $Cu(acac)_2$ sorbed at various experimental conditions. The parameter α^2 is a measure of the covalency of in plane σ bonding between a 3d copper orbital and the ligand orbitals, and β_1^2 is a measure of the covalency of the in plane π-bonding. The π bonding of $Cu(acac)_2$ in the zeolite is more ionic than it is in silica gel and this may be due to a lowering of symmetry resulting from the interaction of the quasi-π-electrons in the complex with the electrostatic field from the cations in the zeolite. With this interpretation, the copper atom is no longer coplanar with the 2acac. The ESR changes occurring with the addition of NH_3 or aniline indicate their coordination as apical ligands.

When pyridine is added (67) the $Cu(acac)_2$ is transformed in the zeolite to $Cu(pyridine)_4^{2+}$. The mechanism proposed for this reaction follows: There is a lowering of the symmetry and stability of the $Cu(acac)_2$ as has been observed before in the zeolite.

Attack by two adsorbed pyridine molecules and two pyridinium ions, which are formed by combining with protons on the acid sites, is thus facilitated to promote this substitution reaction. This reaction has not been observed in ordinary solvents.

E. Concluding Remarks

The aim of the above has been to present pertinent examples of the use of spectral techniques for elucidation of the structure and properties of zeolites. An exhaustive survey of the literature was considered to be unnecessary for this purpose. It is felt that the selective approach employed has facilitated this objective. For example, it has been shown by judicious reference to the literature how infrared spectra in particular provide valuable data for classification of the zeolite's framework and water or hydroxyl linked to it. With this presentation it is quite evident that additional work in this direction with Raman spectra should provide a more quantitative description of the localization and types of cavities than has thus far been obtained.

It has been clearly shown as well by this approach that compilation of visible spectra together with magnetic measurements (ESR, Mössbauer and magnetic susceptibilities), when transition elements occupy zeolite sites, enables the tracing of the ion to its coordination site, and, through comparison with the same ions in aqueous solutions, crystals or minerals, definition of the site, its geometry and type of bonding of the ion to the site. Very valuable quantitative parameters have been shown to be obtained through use of these methods and especially through ESR measurements (degree of covalency, etc). When the results acquired for the selected studies of the transition elements Mn, Fe, Co, Ni and Cu were compared, parallel behavior, according to the type of zeolite, was demonstrated. Geometries and spectral parameters similar to those known in solids or solutions were found in most cases. However, some of the species assumed to be formed in the zeolites were not previously known; e.g., a tetrahedral $Co(H_2O)_4^{2+}$ has been mentioned, though only tentatively (68). Such species are still unknown in solution and it seems unlikely that they exist even in a zeolite. In general, it may be concluded that the species formed is dependent on the geometry of the site. In the case of Ni, Co and Cu a trigonal symmetry (D_{3h}) is implied to exist in dehydrated zeolites which have trigonal sites. The spectral bands are assigned to the transitions in this energy level scheme and ESR parameters are assigned. Until recently, no such complexes were known. However, since the late sixties extensive work has been done on five coordinated D_{3h}, trigonal bipyramid complexes of cobalt, nickel and copper (69-72). The complexes studied were generally with different ligands and it seems that no complex is known in which all five ligands are oxygens, as in the case of zeolites. As for Ni^{2+}, most of the known complexes are low spin, and it would be interesting to perform magnetic susceptibility measurements on Ni^{2+} sorbed on zeolites.

The adsorption of gases on these transition element ions, widely studied by ESR and spectral measurements, has also been examined by selected sampling of the literature to note the surprising stabilizing effect of the zeolite on the oxidation-reduction properties of these ions. Cobalt, for instance, maintains divalency while adsorbed N_2O desorbs at 150°C and decomposes into N_2 and O_2. There are certainly some catalytic effects of the cobalt ion which should be studied in further detail. The oxidation-reduction properties of the zeolite towards iron ions have not as yet been studied in detail and are not well understood. Finally, a glance at the properties of copper bisacetylacetonate (66,67) sorbed on a zeolite and the adsorption of different molecules with the formation of either an adduct or the complete substistitution of the (acac) reveals the wide range of possibilities for further study that these techniques provide.

IV. ORGANIC ION EXCHANGERS

A. General Properties

Organic ion exchangers are composed of a matrix - a highly polymerized three-dimensional network of carbon-hydrogen chains. Charged groups incorporated in this matrix determine the ion-exchange properties of the resultant gel. The negatively charged functional units attached to the matrix may be $-SO_3^-$, $-COO^-$, $-PO_3^{2-}$, etc. in the cation exchanger, while the following cations NH_3^+, NH_2^+, N^+ may be bound to the matrix of the anion exchanger. The matrix of the exchanger is, in general, hydrophobic, water insoluble and does not swell. Through the introduction of the fixed ions, hydrophilic properties are introduced. The ion-exchange resins are thus cross-linked, insoluble polyelectrolyte gels with a somewhat limited capacity for swelling. In contrast to the zeolite structure, there is no uniform periodic structure and therefore no uniform pore size. The behavior of these types of exchangers depends on the cross-linking of the matrix and the nature of the fixed ion. The number of hydrophilic groups and the degree of cross-linking determine, among other characteristics, swelling of the resin, mobility of the ions and, thus, the rate of exchange.

The structure of the matrix determines the thermal and chemical stability of the resin. The nature of the fixed ions determines the ion exchange properties, i.e. a $-SO_3^-$ ion acts as a strongly acidic anion and may be used over a wide pH range, while a $-COO^-$ ion will give in acid solution an undissociated -COOH group and will be capable of ion-exchange reactions only in weakly acidic, neutral or basic solutions. Many models have been proposed for the ion-exchange process, and in all of them hydration of the ions is of great importance.

Spectral methods, especially IR and NMR (21,21a,73-83), have been employed to elucidate ion-exchanger properties. Zundel et al. (21,21a,73) prepared thin films of polyelectrolytes to study primarily ion-water interaction (see section on membranes). Cohen and Peretz (74) studied the water present in cation and anion exchangers for a series of counterions by using the near infrared water band at 1.9 μ. The shift of this band was related to the hydration number of the various ions.

In a paper by Creekmore and Reilley (75), the chemical shift difference between the resin phase water and the bulk water (in an aqueous suspension of ion exchange resins) was used to evaluate the hydration number of a series of counterions in the resin phase. Figure 12 shows that the effect of cations on the proton shift of water in a homogeneous electrolyte solution and in an ion-exchange resin is quite similar. Ions identified as structure-breakers and

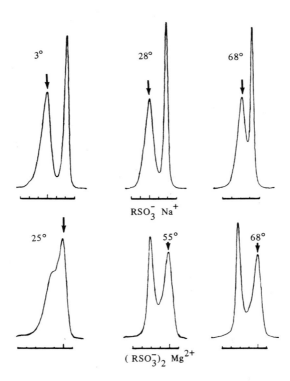

Fig. 12. Comparison of effect of cations on proton shift of water in the case of a homogeneous electrolyte A and an ion exchange resin B (shifts are normalized to 1M) (75).

-formers in solution retain these respective properties in the resin systems as well. The chemical shifts in the ion exchangers are temperature dependent and Figure 13 shows that with an increase in temperature the distance between the two water peaks (in a magnesium resin) becomes greater. This is due to differences in the effect of temperature on the chemical shift of external and internal water. From this temperature dependence, hydration numbers in the resin were evaluated for a series of ions and compared to those of the ions in electrolyte solutions (Tables IV and V). It may be seen from these tables that the same trends are found for both systems. In general, the hydration numbers of cations in the resin are lower than those found for cations in electrolyte solution by this method. The hydration numbers in anion exchangers are approximately one and are consistent with the

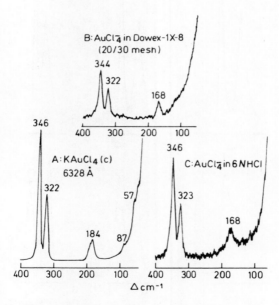

Fig. 13. The effect of temperature on the proton chemical shifts for a sodium and magnesium Dowex 50W-8X resin. The two peaks correspond to the interior and exterior water, the latter being indicated by an arrow (each division 10 Hz) (75).

Table IV

A. Hydration Numbers of Various Cation Resins, Dowex 50w-X8 (50-100 Mesh Resins)(75)

Counterion	Molality	$d\Delta/dT$	h
H^+	4.4	0.00242 ± 0.00011	2.9
Na^+	5.4	0.00261 ± 0.00008	2.9
K^+	6.3	0.00297 ± 0.00011	2.7
Rb^+	5.9	0.00278 ± 0.00010	2.6
Cs^+	5.8	0.00318 ± 0.00004	3.2
Mg^{2+}	2.65	0.00305 ± 0.00015	6.7

B. Hydration Numbers for Anion Exchange Resins, Dowex 1-X8 (50-100 Mesh)

Br^-	5.8	0.00127 ± 0.00023	1.3
I^-	8.9	0.00178 ± 0.00003	1.2

hydration numbers of $(CH_3)_4NCl$ and $(CH_3)_4NBr$ in solution.

The effect of crosslinking on hydration has also been studied (75). The hydration numbers increase with a decrease in the crosslinking, i.e. as the resin phase becomes more dilute.

We would like to mention briefly, at this point, a most recent approach to the resolution of hydration parameters in Dowex 50 and its polyelectrolyte analogue that has been developed by Marinsky and Högfeldt(84). It is based on a semiempirical analysis of the colligative properties of these systems. Though not based on spectroscopic measurements, the primary thrust of this chapter, their approach is unique and

Table V.

Hydration Numbers of Electrolytes Determined by
Temperature Variation Method (75)

Electrolyte	Molality	h
HCl	-	3.4
NaCl	3.08	4.6
NaBr	3.01	4.4
NaClO$_4$	2.98	3.0
Na (p-Toluene	1.56	3.5
(sulfonate	2.47	3.0
	2.73	2.8
	3.29	2.9
KF	2.79	4.4
KCl	2.99	4.6
RbCl	3.20	4.0
CsCl	3.02	3.9
MgCl$_2$	2.92	8.2
(CH$_3$)$_4$NCl	2.66	0.6-1.0
(CH$_3$)$_4$NBr	2.92	1.0

interesting. It leads to hydration parameters in qualitative agreement with those resolved from the chemical shift data.

B. Complex Species Sorbed on Cation and Anion Exchange Resins

Additional insight with respect to the ion-exchange process can be derived from information obtained concerning the structure of the sorbed ion. Such data should, by providing the correct formulation of chemical reactions encountered in the exchange process, enable the quantitative mass-action expression of such equilibria. Spectral methods are particularly suitable for determining the chemical form of the ions in the network of an ion exchangers. A comprehensive survey of the general work in the field of metal complexes that is related to ion exchange and solvent extraction has been compiled and critically reviewed by Marcus and Kertes (85,85a). Faber and Rogers (50) were among the first to measure the ESR spectra of manganese, copper and oxovanadium ions sorbed on cation- and anion- exchange resins. They found that

SPECTROSCOPIC STUDIES OF ION EXCHANGERS

the type of bonding of the species sorbed on cation exchangers is essentially ionic, while in the anion-exchange resins the bonding is covalent. Quantitative parameters such as $g_\parallel$, $g_\perp$, A and B were evaluated.

Spectral measurements in the visible regions were performed for the study of chlorocomplexes in liquid-anion exchangers by Lindenbaum and Boyd (86). The spectra of Fe^{3+}, Co^{2+}, Cu^{2+} and Ni^{2+} in the liquid-anion exchanger were found to be identical to those of the corresponding tetrahedral species, MCl_4^- and MCl_4^{2-}, although none of these species exist in the aqueous solution from which these complexes were extracted by the exchanger.

Nortia and Laitinen (9,10,12) have studied the visible and near-infrared spectra of divalent nickel, cobalt and copper ions sorbed on a cation exchanger with sulfonic and carboxylic functional groups. The spectra of ions sorbed on the moist sulfonic exchanger were identical to their spectra in aqueous solution. Upon drying of the resin, the shift in the absorption bands that was obtained was attributed to the interaction of the ion with the sulfonic group. From the spectral bands obtained for the ions sorbed on a carboxylic resin, these workers assumed that there is at least a partial coordination of the ions by the carboxylic group. In another study the same authors (12) found that the ions sorbed on the cation exchange resins react with ammonia and aminopolycarboxylate solutions. The spectra measured showed that the water molecules in the coordination sphere are replaced by these ligands. The number of molecules of water replaced depended on the concentration of the ligand solution.

Infrared spectroscopy is not a suitable method for spectral study of ion-exchange resins since the organic network absorbs strongly in many regions. In the far infrared region, where most of the metal-ligand vibrations are found, measurement is almost impossible. However, as will be seen below, it may be used in a supportive role in many cases. With the development of gas lasers, Raman spectra are resolvable in exchangers (22,23) and some halide

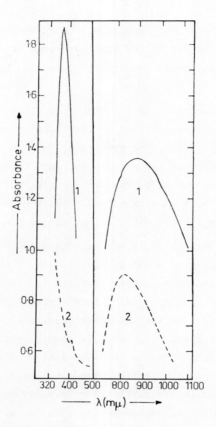

Fig. 14. Raman spectra of $AuCl_4^-$. (23)

complexes sorbed on anion exchangers have been examined with this technique. The chloro- and bromo-complexes of gold sorbed on an anion-exchange resin have been studied, and their spectra compared to the Raman spectra of the corresponding solids and the species present in the equilibrating acid solution (Fig. 14). These ions, $AuCl_4^-$ and $AuBr_4^-$, were found to have D_{4h} symmetry. Complex species of indium chloride, bromide and iodide have been studied as well (22,23). These complexes are known to be tetrahedral, however, the Raman spectra were ambiguous. The octahedral $ReCl_6^{2-}$ has been chosen for Raman study as well; however, the results obtained were complicated by partial hydrolysis of this ion in the resin.

Table VI

Isomer Shift (87)

Sample	Temperature	Isomer shift mm/sec	Coordination
Wet resins	liq. N_2	0.57	
Dry resins	liq. N_2	0.54	
$KFeCl_4$	liq. N_2	0.54	tetrahedral $4Cl^-$
$KFeCl_4$	room	0.54	tetrahedral $4Cl^-$
NH_4FeCl_4	room	0.50	tetrahedral $4Cl^-$
$FeCl_3$	liq. N_2	0.73	octahedral $6Cl^-$
$CeCl_3$	room	0.70	octahedral $6Cl^-$
$FeCl_3 \cdot 6H_2O$	room	1.12 0.12	octahedral

Iron ions in ion exchange resins have been studied by Mössbauer spectroscopy (26,87,88) and magnetic susceptibility measurements (89). The Mössbauer spectrum of ferrous ions sorbed on Dowex 50 is characteristic of high spin Fe^{2+}; the spectrum area is greatly enhanced by a decrease in temperature or by dehydration (26). These results show that the iron ions are solvated in the water-swollen resin. Ferric ions were sorbed from hydrochloric acid solutions on an anion exchange resin (87). A single sharp absorption peak was found in all the Mössbauer spectra, indicating a symmetrical charge distribution around the iron atom. The isomer shift obtained is compared with isomer shifts reported for a number of solid ferric compounds in Table VI. Its value is closest to that observed for the tetrahedral chlorocompounds and it may be concluded that the ferric ion is sorbed as the tetrahedral chlorospecies, $FeCl_4^-$. The small difference between the isomer shift of the wet and the dry resin samples indicates that no water molecules are in the coordination sphere.

Rutner (1) has measured the reflectance spectrum of ferric ions sorbed on an anion-exchange resin from hydrochloric acid solutions. The absorption band at 362 mµ was assigned to $FeCl_4^-$. Johansson (88) recorded Mössbauer spectra for ferric and ferrous ions sorbed on cation exchange resins with varying degrees of crosslinking. The absorption lines obtained were about three times as wide as those found in crystalline substances to show that broadening is connected with variations in the value of the electric field gradient tensor for the different cationic sites in the noncrystalline lattice. Hydration of the resin was found to cause a decrease in the area of the absorption curve, thus confirming an earlier observation of this effect (26). This result has been interpreted as a weakening of the attractive force between the ions of opposite charge that arises from solvation of the ion.

Nortia and Kontas (89) have measured the magnetic susceptibility of the aquo iron(III) dimer sorbed on Dowex 50. The iron(III) system is very complex due to hydrolysis, and such determination of magnetic moments in the exchange resin where no other approach to the examination of species sorbed by the resin has been employed is believed to be inconclusive for this purpose. Indeed, few magnetic susceptibility measurements of ion-exchange resins and zeolites have so far been reported (54).

Another use of Mössbauer spectroscopy was recently made by Stein and Marinsky (90). The relative amount of monomeric and dimeric anion present in ferric salts of EDTA, HEDTA, CDTA and NTA in solid, solution, polyelectrolyte and ion exchanger was studied as a function of pH. In general Mössbauer spectral parameters were similar for polyelectrolyte and resin samples when the equilibrating solutions were adjusted to the same pH. The quadruple splitting (QS) value characteristic for the dimer in the solution for each salt was also found in the polymeric phase. This suggests that the size of the cation has no influence on the QS parameter and that the interaction between the cation and the ferric polyamino-carboxylate is purely electrostatic. This conclusion is in agreement with those deduced from the earlier spectral results (see concluding remarks). In the chelating resin Dowex A-1 only monomeric anions were found. This was expected since the functional groups, bound to an inflexible resin matrix, cannot assume the appropriate geometry for dimerization to occur.

Study of the stepwise formation of complexes by the use of ion exchangers has been performed in our laboratory. Regarding formation of a complex MA_n, the only data generally available on the intermediate species are the individual stability constants. As n increases in these complexes, techniques for evaluation of the stability constants become more complicated and less reliable. In solution there is a mixture of different complex species rendering the interpretation of spectral results almost impossible. An attempt was therefore made to obtain a selective sorption of the various species formed in the stepwise reaction and then to define each species by measurement and interpretation of its spectrum. The spectra obtained were compared to those found in solution (aqueous and nonaqueous) or in the solid. In most cases, the initial species (solvated) and the final one (MA_n) have been defined in the literature. Assignments of the spectra and evaluation of quantitative parameters enable the determination of the structure of the species sorbed. Selective sorption of one or another species on the exchanger may be controlled by a) changes in the equilibrating solution and by b) use of different ion exchangers.

Copper ions were sorbed on cation- and anion-exchange resins from aqueous chloride solutions and the absorption spectra were measured (14) (Figs. 15,16,17). The species sorbed on the cation-exchange resin from a dilute copper chloride solution has a single absorption band in the visible region at 12,200 cm^{-1} ($\varepsilon_{mol} = 11$). This absorption band corresponds to the octahedral $Cu(H_2O)_6^{2+}$. If the equilibrating solution contains an excess of chloride ions, the absorption maximum is shifted to lower energies, 11,400 cm^{-1}, while the molar absorption coefficient increases to about 60. This shift may be related to a replacement of some water molecules by chloride ions while the increase in intensity of the band may be

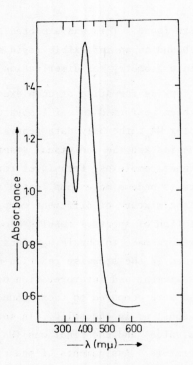

Fig. 15. 1) $CuCl_2$ sorbed on Dowex 50; 2) $Cu(NO_3)_2$ sorbed on Dowex 50. (14)

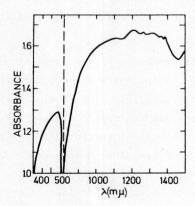

Fig. 16. Solution of 0.05M $CuCl_2$-KCl sorbed on Dowex 1. (14)

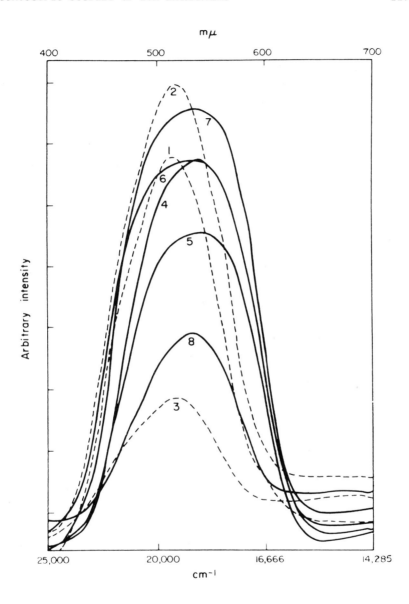

Fig. 17. Solutions of 0.5M $CuCl_2$ sorbed on Dowex 1. [Absorbance measurements between 350 and 500 mμ were made using an attenuator (optical density 2.0) in the reference beam.](14)

connected to a decrease in symmetry of the coordination shell making the d-d transition less forbidden. Analysis of the species sorbed on the cation exchanger showed that the ratio of Cu:Cl was about 1:1. It was therefore assumed that the species sorbed on the cation exchanger is $[CuCl(H_2O)_5]^+$. Two different spectra were measured on the anion exchanger, the type of spectrum depending on the chloride/copper ratio in the equilibrating solution.

The absorption bands obtained at a high Cl:Cu ratio are also found in non-aqueous solvents and in single crystals and have been assigned to the tetrahedral $CuCl_4^{2-}$ (91,92). The same species is sorbed on the anion-exchange resin. The spectral bands of the species sorbed from solutions containing low chloride/copper ratios were not as easily interpreted since the assignments in the literature are ambiguous.

Manahan and Iwamoto (93) assigned the band at 20,840 cm^{-1} to the copper species, $CuCl_3^-$, in acetonitrile. In the anion-exchange resin this band appears together with a broad tetrahedral band in the absorption region between 7400 and 10,000 cm^{-1}, which may be due to an unsymmetrical tetrahedral $CuCl_3H_2O^-$ species. With this interpretation of these results the species, $CuCl_4^{2-}$ and $CuCl_3H_2O^-$, are presumed to be sorbed together. A mixture of $CuCl_3^-$ and $CuCl_4^{2-}$ might explain the broad band in the 20,000-25,000 cm^{-1} region as well. Such a mixture could also explain the distribution measurements obtained recently by Waki et al. (13). From their spectral measurements (which are identical to those presented by the above authors) Waki et al. (13) conclude, however, that a dimeric species $Cu_2Cl_7^{3-}$, is sorbed.

Electron spin resonance measurements have also been made of the various species sorbed. The results of these measurements are shown in Table VII. From the absorption and ESR spectra, λ, the spin orbit coupling constant, has been calculated. These λ values indicate qualitatively the extent of covalent bonding in each configuration; i.e. a decrease in λ may be understood as an increase in charge transfer from the ligand to the central ion.

Table VII

Electron Spin Resonance Measurements (14)

Resin	Species sorbed	Δ cm^{-1}	Line widths gauss	g (av)	λ cm^{-1}
Dowex 50	$Cu(H_2O)_6^{2+}$	12,100	110	2.192	575
Dowex 50	$CuCl(H_2O)_5^{+}$	11,400	130	2.178	502
Dowex 1	$CuCl_3H_2O^{-}$	8,300	340	2.130	270
Dowex 1	$CuCl_4^{2-}$	8,300	360	2.124	251

In a parallel cobalt chloride study (94,95), it was possible to identify the following species: on the cation exchanger - $Co(H_2O)_6^{2+}$, $Co(H_2O)_5Cl^{+}$, $Co(H_2O)_3Cl^{+}$ (tetrahedral) and on the anion exchanger - $CoCl_3H_2O^{-}$, $CoCl_4^{2-}$.

Cobalt chloride species were also sorbed onto ion-exchange resins from non-aqueous solutions (ethanol and acetone). From the small shift of the absorption bands, it may be implied that water is replaced by ethanol or acetone in the species sorbed.

Among the other cobalt systems studied (96,97) the nitrate (94) is the most interesting. The nitrate ion, a weak complexing agent for transition ions, is not known to complex cobalt in aqueous solutions. Addison and Gatehouse (98) have prepared an anhydrous tetranitrato cobaltous complex starting with the metal and dinitrogen tetraoxide in organic solvent mixtures. This compound has been found to be one of the rare complexes of cobalt with a coordination number of eight, each nitrate acting as a bidentate ligand. The absorption spectra of cobalt in aqueous and nonaqueous solutions (ethanol, acetone or acetonitrile) show no indication whatever of formation of a nitrato compound.

Table VIII lists the spectral bands of cobalt species sorbed from nitrate solutions on a cation- and anion- exchange resin.

Table VIII

Absorption Bands of Cobalt Ions in cm^{-1} Sorbed on Cation and Anion Exchangers (94)

Resin	Solvent	Equilibrating media	Near i.r. 1	Vis. 2	U.V. 3	$f_1^* \times 10^3$	$f_2^* \times 10^3$
Dowex 50-Li^+	Water	0.1M $Co(NO_3)O_2$	6990,7695(3,7)†	19,300(12)	37,750;31,200	0.04	0.19
Dowex 50-Li^+	Water	0.1M $Co(NO_3)_2$, 2M $NaNO_3$	8000(11)	19,300(33)	37,750;31,200	0.11	0.50
Dowex 1-NO_3^-	Water	1M $Co(NO_3)_2$	8475(26.4)	18,570(247)	30,300	0.41	4.30
Dowex 1-NO_3^-	Water	1M $Co(NO_3)_2$, 9M $NaNO_3$	8475(53.2)	18,520(340)	29,950	0.79	5.34
Dowex 1-NO_3^-	Ethanol	1M $Co(NO_3)_2$	8475(97.6)	18,520(406)	30,330	1.59	8.2
Dowex 1-NO_3^-	Ethanol	1M $Co(NO_3)_2$, sat.$NaNO_3$	8475(87.4)	18,520(334)	29,950	1.29	7.90
Dowex 1-NO_3^-	Acetonitrile	0.05M $Co(NO_3)_2$	8475(63.4)	18,520(343)	30,120	0.94	7.00
Dowex 1-NO_3^-	Acetonitrile	0.05M $Co(NO_3)_2$, sat.$NaNO_3$	8475(13.7)	18,650(162)	29,950	0.16	2.86
Dowex 1-NO_3^-	Acetone	1M $Co(NO_3)_2$	8400	18,720	29,950		
Dowex 1-NO_3^-	Acetone	1M $Co(NO_3)_2$, sat.$NaNO_3$	8340	18,870	29,950		

*f = oscillator strength.
†Molar extinction coefficients given in parentheses.

Comparison of the spectra of species sorbed on the cation-exchange resin in the absence and presence of an excess of nitrate ion in the equilibrium solution shows a very small red shift of the near-infrared band and an increase in ε_{mol} from 12 to 33 for the species sorbed from solutions containing an excess of nitrate ion. This may indicate the sorption of a cationic octahedral mononitrate species.

Spectra of the species sorbed on cation and anion exchangers are shown in Figs. 18 and 19. The species sorbed on the anion exchanger from the nitrate solution shows a striking change in color (i.e. deep magenta) and, consequently, a different spectrum. An asymmetric broad peak between 18,870 and 18,450 cm^{-1} reaches ε_{mol} values of 400, while a second peak is situated between 8340 and 8475 cm^{-1}.

The infrared spectrum of the sorbed species was measured and found to be different than the nitrate spectrum obtained with a resin sample equilibrated in the nitrate solution free of cobalt ions. The new peaks which appear in the spectrum of the sorbed cobalt nitrate species (Fig. 20) are at 1295, 1025, 813 and 1460-1480 cm^{-1}, and indicate a decrease in symmetry of the free nitrate ion (D_{3h}) to a nitrato ion $O-NO_2^-$ (C_{2v}) in the cobalt complex. The absorption bands, ε_{mol}, oscillator strength and infrared spectrum indicate the sorption of a tetrahedral $Co(ONO_2)_4^{2-}$ species similar to the compound previously described (98). The crystal field splitting (10Dq) and nephelauxetic [B] parameters were evaluated (15) and found to agree well with those for the solid compound [10 Dq = 4775 cm^{-1} (vs 4660 cm^{-1}) and B = 818 cm^{-1} (vs 855 cm^{-1})].

The complex species of cobalt-thiocyanate were also studied by this technique (97). The following species were characterized by their sorption on the ion-exchange resins and measurement of their absorption spectra: $[Co(NCS)(H_2O)_5]^+$, $CoNCS(H_2O)_3^+$, $Co(NCS)_3(H_2O)^-$ and $Co(NCS)_4^{2-}$]. Table IX presents the spectral parameters computed for this system. The 10 Dq and B values evaluated for the

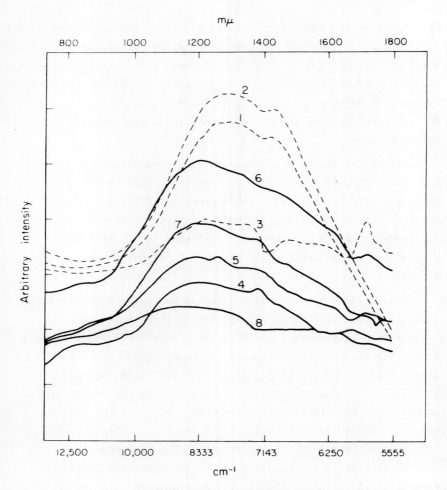

Fig. 18. Visible absorption spectra of cobalt(II) species sorbed on ion exchangers from aqueous and nonaqueous solutions.
1. Dowex 50WX8Li$^+$, 1.0M Co(NO$_3$)$_2$ in water;
2. Dowex 50WX8Li$^+$, 2.0M Co(NO$_3$)$_2$ in water;
3. Dowex 50WX8Li$^+$, 1.0M Co(NO$_3$)$_2$, 2M NaNO$_3$ in water;
4. Dowex 1X8NO$_3^-$, 1.0M Co(NO$_3$)$_2$, 9M NaNO$_3$ in water;
5. Dowex 1X8NO$_3^-$, 0.5M Co(NO$_3$)$_2$ in ethanol;
6. Dowex 1X 8NO$_3^-$, 1.0M Co(NO$_3$)$_2$ in acetone;
7. Dowex 1X8NO$_3^-$, 0.1M Co(NO$_3$)$_2$ in acetonitrile;
8. Dowex 1X8NO$_3^-$, 0.05M Co(NO$_3$)$_2$, sat. NaNO$_3$ in acetonitrile. (96)

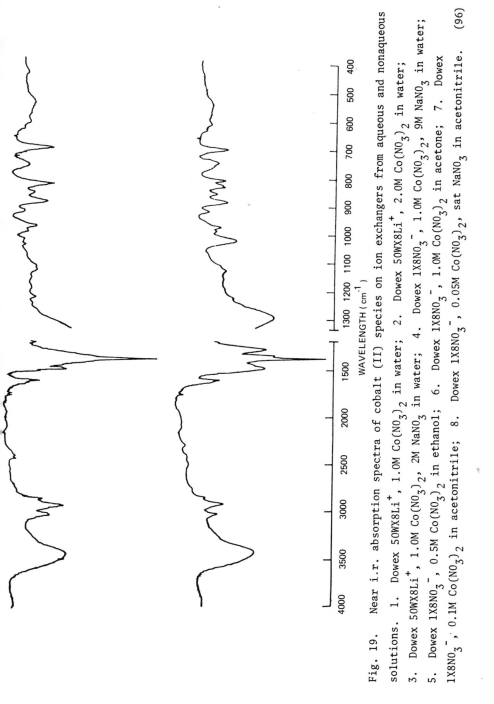

Fig. 19. Near i.r. absorption spectra of cobalt (II) species on ion exchangers from aqueous and nonaqueous solutions. 1. Dowex 50WX8Li$^+$, 1.0M Co(NO$_3$)$_2$ in water; 2. Dowex 50WX8Li$^+$, 2.0M Co(NO$_3$)$_2$ in water; 3. Dowex 50WX8Li$^+$, 1.0M Co(NO$_3$)$_2$, 2M NaNO$_3$ in water; 4. Dowex 1X8NO$_3^-$, 1.0M Co(NO$_3$)$_2$, 9M NaNO$_3$ in water; 5. Dowex 1X8NO$_3^-$, 0.5M Co(NO$_3$)$_2$ in ethanol; 6. Dowex 1X8NO$_3^-$, 1.0M Co(NO$_3$)$_2$ in acetone; 7. Dowex 1X8NO$_3^-$, 0.1M Co(NO$_3$)$_2$ in acetonitrile; 8. Dowex 1X8NO$_3^-$, 0.05M Co(NO$_3$)$_2$, sat NaNO$_3$ in acetonitrile. (96)

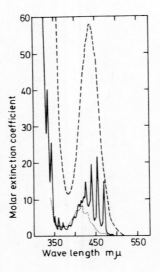

Fig. 20. Infrared spectra of Dowex 1 - NO_3^- (upper curve) and and Dowex 1 - $Co(ONO_2)_4^{2-}$ (lower curve) (96a).

Table IX

Spectroscopic Parameters for Sorbed Thiocyanate Species (97)

Species	$f_1 \times 10^3$ near i.r.	$f_2 \times 10^3$ visible	10 Dq(cm^{-1})	B
$Co(NCS)(H_2O)_5^+$	0.30	0.97	9390	854
$CoNCS(H_2O)_3^+$	4.34	12.60	3860	740
$Co(NCS)_3H_2O^-$	6.47	16.32	4180	771
$Co(NCS)_4^{2-}$	6.72	20.85	4360	688

intermediate species from the spectra are in good agreement with those obtained by using "the rule of the average environment."(99) Infrared measurements were also made in order to ascertain the type of bonding in the complex species sorbed. In the infrared spectra of cobalt sorbed from thiocyanate solution on an anion-exchange resin bands were found for cobalt bound through nitrogen (2060,

860 cm^{-1}) and through sulphur (2150,728,710 cm^{-1}) in the same sample. It is generally accepted that thiocyanate is bound to the cobalt ion through the nitrogen (100,101). However, as the matrix of the exchanger is very compact, different modes of bonding seem possible. In some cases weak bonds with the sulphur may be formed (102). The tertiary amine group in the anion exchanger may be electropositive enough to enable the coordination and formation of an asymmetric bridge: $-N-(CH_3)_3-NCS-Co$ or $-N(CH_3)_3-SCN-Co$ to explain the observed result.

From the quantitative parameters obtained for the various cobalt species, it was possible to place the ligands in the following series (15):

The spectrochemical series (103): $Cl^- < H_2O < NCS^- < NO_3^-$

10 Dq(cm^{-1})   3219   3800   4360   4775

The nephelauxetic series (104): $H_2O > NO_3^- > Cl^- > NCS^-$

B(cm^{-1})   826   818   753   725

Good agreement was found with earlier compilations of these series (103,104), the only inversion being that between the B values of NCS^- and Cl^-. As the entire difference between these two values is less than 5%, it may well be within the limit of error (for a d^7 system), or it is possible that this inversion is due to the somewhat different type of bonding of the thiocyanate to the cobalt in the anion-exchange resin (97).

The uranyl complex species with sulfate (105), thiocyanate (106), nitrate (106), fluoride, chloride and bromide (107), have also been studied. The uranyl group OUO^{2+} is linear (symmetry $D_{\infty h}$) and is not greatly affected by complexing agents. The most common coordination number of uranyl is six and the complex formed may be represented as a hexagonal bipyramid in which the six monodentate or the three bidentate ligands are in the same plane as the uranium ion, the two oxygens one above and the other below this plane. The spectrum of uranyl ion between 330 and 550 mμ is rather complicated due to the vibrational coupling of the electronic transitions. Upon complexation, relatively small energy shifts and changes in the vibrational pattern and in the molar absorption values occur (108-9). The following species were found to be sorbed from sulfuric acid or lithium sulfate solutions: (105)

On Dowex 50: $UO_2(H_2O)_6^{2+}$ and $UO_2(HSO_4)^+$

On Dowex 1 : $UO_2(SO_4)_2^{2-}$ and $UO_2(SO_4)_3^{4-}$.

Until now, there has been no evidence for the existence of a trisulfate species in aqueous solutions. From the IR spectra, a splitting of the sulfate bands has been observed to indicate a decrease in symmetry from the free sulfate group $[T_d]$ to the bound sulfato group $[C_{2v}]$. Finally, upon replacing water molecules by sulfate ions in the coordination shell, the asymmetric stretching frequency of the UO_2^{2+}, ν_3, decreases from 958 to 905 cm^{-1}, i.e. the U-O bond distance increases from 1.719 to 1.74 Å. A very small decrease in bond order was observed.

On studying the nitrato-uranyl system (106), it was found that the $UO_2NO_3^+$ and $UO_2(NO_3)_3^-$ species are sorbed. They were identified through their spectra on the respective ion-exchange resins. In an anion exchange study made earlier by Ryan (8), it was assumed that the species sorbed was $UO_2(NO_3)_4^{2-}$ together with some $UO_2(NO_3)_3^-$. While the $UO_2(ONO_2)_3^-$ species and its spectrum are well known from solids and nonaqueous solutions, it is very difficult to synthesize the tetranitrato species. This compound apparently has been prepared only by fusion (8). This type of preparation interferes with analysis of the new compound, and its composition is doubtful. (The difficulty in preparing such a species probably arises from the fact that while in the trinitrato complex, all the nitrate groups are bidentately linked, the tetranitrate cannot have more than two nitrates bidentate with the other mondentate. Thus, bond breaking has to take place for its formation.) The spectrum obtained for this compound is completely different from that of the trinitrato complex and is completely structureless (Fig. 21); the spectrum of the trinitrato

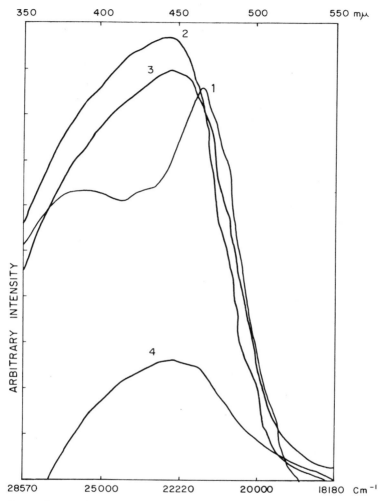

Fig. 21. Absorption spectra of solid complex U(VI) nitrate salts compared to that of the aqueous uranyl ion: --- [$(C_2H_5)_4N$][$UO_2(NO_3)_4$]; ___ $(C_2H_5)_4NUO_2(NO_3)_3$; ... uranyl nitrate in 0.514M HNO_3[8].

complex exhibits a strong vibrational interaction in the lower energy region.

Spectral study of resin-sorbed halide complexes of uranium have led to the identification of the following species (107): UO_2F^+, $UO_2F_3^-$, UO_2Cl^+, $UO_2Cl_3^-$, $UO_2Cl_4^{2-}$, UO_2Br^+ and $UO_2Br_4^{2-}$. The uranyl thiocyanate complex species (104) have also been studied, however interpretation of their visible spectra is complicated for two reasons: a) no spectra are currently known in solution or in solids because of a low energy charge transfer band dominating this region; b) the absorption spectra of the species sorbed on a cation and anion exchanger exhibit definite absorption bands. However, almost no vibrational interaction has been observed (Fig. 22). IR spectra show that the thiocyanate group is linked to the uranium through the nitrogen atom. From these results it has been assumed that UO_2SCN^+ and $UO_2(SCN)_4^{2-}$ are sorbed on the ion-exchange resins.

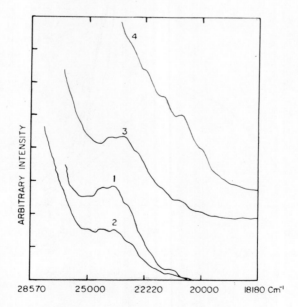

Fig. 22. Visible spectra of uranyl thiocyanate species:
1 - from 2N KSCN solutions on Dowex 50;
2 - from 2N KSCN on Dowex 1 - Cl;
3 - from 6N KSCN on Dowex 1 - Cl;
4 - from 10N KSCN on Dowex 1 - Cl.   (106)

A most important result of the research on uranyl species which has been obtained through comparison of the spectra measured on the various ion exchangers (110) is the observation that species with the same symmetry, independent of the type of ligand, present very similar spectra, i.e. vibrational patterns. The strength of the ligand, deduced from the spectrochemical series, causes only a small blue or red shift of the spectral curve. From Fig. 23, it can be seen that all the cationic species of the type UO_2X^+ have very similar spectra which are also very similar to that of the uranyl hexahydrate. It may be thus inferred that these species are $UO_2X(H_2O)_5^+$, obtained by replacing one molecule of water by a

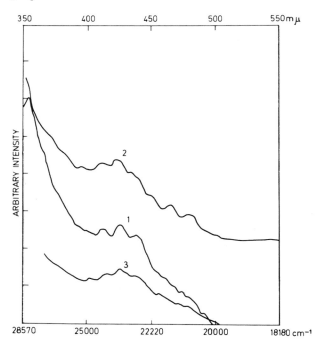

Fig. 23. Visible spectra of cationic uranyl species:
1 - $UO_2NO_3^+$;
2 - $UO_2HSO_4^+$;
3 - UO_2Cl^+;
4 - UO_2Br^+. (110)

ligand X. The species thus obtained is only slightly distorted from the hexagonal bipyramidal form of the hexahydrated species, as the field strengths of the ligands are quite similar to that of water.

Figure 24 shows the spectra of the coordinated species $UO_2X_3^-$ (where X is monodentate) of symmetry D_{3h} and Fig. 25, the tetra-coordinated species $UO_2X_4^{2-}$ of symmetry D_{4h}. In these spectra, the first and second regions of the spectrum are strongly perturbed vibrationally, while the molar extinction coefficient in the second region is much higher than that in the first. In Fig. 26 the spectra of the species $UO_2(SO_4)_3^{4-}$ and $UO_2(NO_3)_3^-$ are

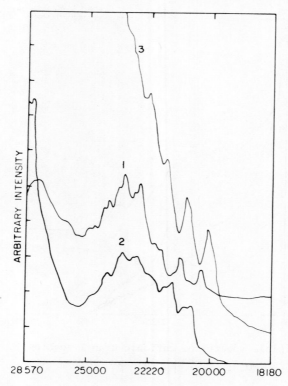

Fig. 24. Visible spectra of anionic uranyl species symmetry D_{3h} (monodentate ligands): 1 - $UO_2Cl_3^-$; 2 - $UO_2F_3^-$; 3 - $UO_2(HSO_4)_3^-$. (106)

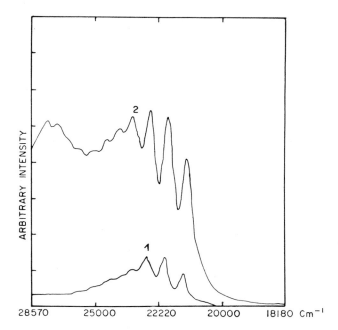

Fig. 25. Visible spectra of anionic uranyl species symmetry D_{4h}: 1 - $UO_2Cl_4^{2-}$; 2 - $UO_2(SO_4)_2^{2-}$; 3 - $UO_2Br_4^{2-}$. (106)

represented. In these complexes all the anions act as bidentate ligands and the coordination number in the equatorial plane of the uranyl group is 6 (symmetry D_{3h}). These results agree with those obtained by Gorller-Walrand and DeJaegere (111) in 1972 for a number of uranyl complexes. The correlation obtained between the symmetry of the species and its spectrum seems to provide a valid criterion for the elucidation of the symmetry and coordination number of complexed uranium species through their absorption spectra.

C. Complex Species Sorbed on Chelating Ion Exchangers

The advantage of using chelating resins in separation chemistry because of their ion specific properties has been significantly diminished by their very slow exchange with ions in solution. Kinetic studies have shown that in the iminodiacetic acid

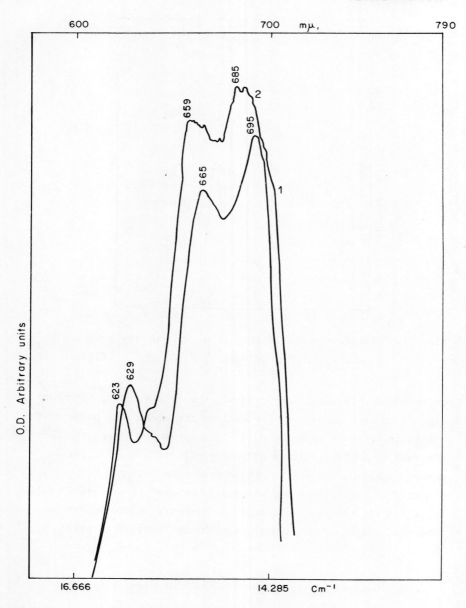

Fig. 26. Visible spectra of anionic uranyl species symmetry D_{3h} bidentate ligands: 1 - $UO_2(NO_3)_3^-$; 2 - $UO_2(SO_4)_3^{4-}$. (106)

resin, the slow step of the exchange reaction for divalent ions is diffusion through the particle (112,113); in the phosphonic acid resin the rate-controlling step for the exchange reaction of uranyl and tervalent transition metal ions is the chemical reaction at the exchange site (114,115), i.e. the making or breaking of strong chelate bonds. Spectral studies can provide insight with respect to the nature of the bonds formed and have been employed to examine this aspect in the carboxylic-, the phosphonic- and the iminodiacetic acid resins. The carboxylic acid resins have been extensively studied by infrared measurements (116-118); the cations investigated were the uranyl, indium and mercury ions.

The presence of two bands in the region of the asymmetric stretching vibrations of the carboxylate ion has been claimed to prove the existence of two different types of metal-ion bonding. This conclusion is refuted by research done in our laboratory (119), since these bands are also found to exist in the absence of a chelating ion, i.e. when the carboxylate group is in the hydrogen or sodium form. The visible spectrum obtained for the copper ion on a carboxylic acid resin shows a maximum absorption at 14,300-14,500 cm^{-1} that is characteristic for octahedral copper species. It may thus be assumed that the species bound to a carboxylate ion and four molecules of water is $(H_2O)_4Cu\genfrac{}{}{0pt}{}{OOC-R}{OOC-R}$. Through ESR measurements, it may be seen that α^2 (the degree of covalency) decreases from 0.9 for the ionic species of $Cu(H_2O)_6^{2+}$ sorbed on Dowex 50 to 0.79 for the species sorbed on the carboxylic acid resin ($\alpha^2 = 0.5$ for a pure covalent bond).

Umezava and Yamabe (120) have studied the ESR spectra of Cu^{2+} in a series of cation exchange resins. They have proposed the formation of the ion $R-C\genfrac{}{}{0pt}{}{O}{O}Cu\genfrac{}{}{0pt}{}{O}{O}C-R$ in the case of the carboxylic acid resin. The formation of such an ion seems unlikely because:
a) this complex should be tetracoordinated; this does not agree with the visible spectrum (119) and b) the carboxylate group acts as bidentate ligand; consequently its symmetry should be changed,

followed by a split of the vibrational bands. No such bands appear in the IR spectrum.

Important and extensive study of the binding of metal ions by polyelectrolytes (polymethacrylic (PMA) and polyacrylic (PAA) acids and their salts) in gel and solution form has been performed mainly by Leyte et al. (121-126), Marinsky et al. (127-130) and Weil et al. (131-132). Mandel and Leyte have studied the chelation of copper(II) ions by polymethacrylic acid using potentiometric, viscometric, electrophoretic and spectrophotometric methods (121, 122). These studies were made over a wide range of pH and show the binding of one copper ion by two carboxylic acid groups. The interaction of alkali and alkaline earth metal ions with polymethacrylic and polyacrylic acid salts has been studied by IR spectroscopy (123). The IR results obtained show that the vibrations of the carboxylate groups are not changed by the size or the charge of these ions, i.e. no site binding occurs between these ions and the carboxylate groups. In another study the same authors (124) have studied the changes which occur in the IR, visible and UV spectra and in the magnetic properties of solutions of PMA-Cu(II) as a function of the degree of neutralization. The binding of copper ions by PMA was studied at different f values [f = (equivalent concentration of ionized PMA)/(equivalent concentration of Cu(II)]. At low f-values (1.3) a shoulder at 26,000 cm^{-1} is clearly discernible in the visible spectrum, it disappears at f = 3.3. The magnetic moment depends strongly on f and shows values of 1.91, 1.54 and 1.66/BM respectively at f = 0, 1.3 and 3.0. Lambert-Beer's law is obeyed (1.3 < f < 3.4) in the concentration range of the measurements. The important conclusion reached from these observations is that the ligands attached to a given Cu(II) ion are from one molecule. The spectral band at 26,000 cm^{-1} for the Cu-PMA in the region 1.3 < f < 3 as well as the subnormal magnetic moments are observed in acetate, succinate, glutamate and other systems where dimer structures, i.e. binuclear copper complexes are known to be formed. In PMA the number of C atoms between neighboring carboxylate groups also equals that in the glutarate so that the formation of binuclear copper complexes in PMA at f = 1.3 has been inferred. At f = 3.3 the 26,000 cm^{-1} band disappears, suggesting the dissociation of the dimer. However, from IR data it has been found that the copper ion still binds two carboxylate groups, i.e. only the Cu to Cu bond is broken. On the basis of symmetry considerations assignment of the spectral bands as well as explanation of the intensities of the bands has been made (more or less forbidden transitions are assigned according to the symmetry of the dimeric and monomeric species).

The studies of ion binding in PMA and PAA made by Marinsky's group were based on potentiometric, polarographic and spectral measurements. The stability constant of CuA_2 determined for PAA (127) was found to be in good agreement with that determined for Cu (acetate)$_2$. Further study by this group (128) showed the existence of only one dominant species, an ion pair with Ca(II), Co(II) and Zn(II); the absence of chelates in PAA and PMA was explained by the unfavorable geometry of the fixed array of functional groups in these polymers. In a recent study of this group (129,130) the complexing of nickel(II), cobalt(II) and copper(II) by PMA gel and a linear polyelectrolyte analogue has been examined. The formation of CoA^+ and ZnA^+ has been confirmed and their formation constants have been shown to correspond to the values reported in the literature for the MA^+ (acetate) species. This result is in disagreement with previous reports in the literature (121-124) which according to the present authors is due to an error in free ligand (A^-) evaluation. The complexation of copper in a PMA gel (130) has yielded results similar to those obtained for the linear PMA and PAA (124). In the neutralization range of 0.35-0.80, the complex species is CuA_2. In the neutralization range 0-0.35 Cu-Cu interaction between neighboring CuA_2 entities has been deduced from the visible spectrum, and subnormal magnetic moments.

Weill et al. have mainly used the NMR technique (131,132) to study the interaction of various ions with polyelectrolytes. With this technique questions with respect to a) stoichiometry and constants of binding, b) the characterization of the binding sites and the state of hydration of bound ions, c) the mobility of the counter-ions, d) the state of water in concentrated solution are in principle resolvable.

The authors have studied (131) the chemical shift of water in Co(II) solutions in the presence and absence of the tetramethylammonium salt of several polyelectrolytes. The polyphosphate has been studied in detail showing that the binding of Co(II) is characterized by a complete loss of exchangeable water (no shift in the presence of the polyelectrolyte). The binding of Co is sufficiently high to remove it nearly completely from solution until its initial concentration exceeds fifty percent of the available phosphate groups. Different polyelectrolytes yield different results, e.g. PAA appears to be partially dehydrated while the behavior of carboxy methylcellulose (CMC) of a high degree of substitution is more complex because of the presence of several types of binding sites.

The phosphonic acid resins have also been investigated by ESR and IR measurements (133-135). Through ESR measurements, it has been found that the bonding between this resin and copper ions is essentially electrostatic (120) while IR measurements suggest that at least some coordinative bonds are formed with transition metal ions (134,135). The authors deduce a strong interaction of uranyl ion with two bidentate phosphonic acid groups (135). This difference in bonding may explain the different kinetic behavior of uranyl and tervalent transition metal ions (114-5).

The most widely studied chelating resin until now has been the iminodiacetic exchanger (11,12,106,119,120,136-138). The iminodiacetic group dissociates in different ways (37) depending on the pH of the solution.

$$\underset{\text{I pH=1.2}}{\text{R–N}^+\text{H}\begin{cases}\text{CH}_2\text{COOH}\\\text{CH}_2\text{COOH}\end{cases}} \rightarrow \underset{\text{II pH=4.0}}{\text{R–N}^+\text{H}\begin{cases}\text{CH}_2\text{COO}^-\\\text{CH}_2\text{COOH}\end{cases}} \rightarrow \underset{\text{III pH=7.4}}{\text{R–N}^+\text{H}\begin{cases}\text{CH}_2\text{COO}^-\\\text{CH}_2\text{COO}^-\end{cases}} \rightarrow \underset{\text{IV pH=12.3}}{\text{RN}\begin{cases}\text{CH}_2\text{COO}^-\\\text{CH}_2\text{COO}^-\end{cases}}$$

The pH of dissociation will be even lower than designated when chelation occurs. In concentrated acids, the resin in form I acts as an ion exchanger and the species sorbed should be comparable to those sorbed on Dowex 1. In neutral or alkaline solutions, it should act as a bidentate (form III) and tridentate (form IV) ligand.

The visible spectra and ESR measurements have clearly shown that the chlorocomplexes of copper and cobalt, $CuCl_4^{2-}$ and $CoCl_4^{2-}$ are sorbed by the resin from hydrochloric acid solutions; the spectral and ESR parameters evaluated are identical to those found for the species sorbed on Dowex 1 (119,136). The species formed from neutral solutions are chelated by the carboxylate groups, however no conclusive evidence is found for chelation through the nitrogen as well (119,136,11). Since the complex is assumed to contain one iminodiacetic group per metal ion, it is inferred that the coordination number reaches six through addition of water molecules. However, it has been shown that if sorption is made from solutions containing a rather high concentration of neutral salt, the replacement of water molecules by the anion shifts the spectral curve, the direction and magnitude of the shift depending on the position of the anion in the spectrochemical series (119, 132,12). These changes may sometimes be observed with the naked eye; copper ions sorbed on the iminodiacetic resin color the exchanger a deep blue. When the sorption is effected from chloride solutions the color of the exchanger changes to a dark green (119). ESR measurements of copper ions on the chelating resin have shown that the bonding is partially ionic ($\alpha^2 = 0.86$). This is not surprising since the α^2 value determined for EDTA is 0.84.

A somewhat more detailed study has been made of uranyl species sorbed on Dowex A-1 (106). In acid solutions the anionic sulfate, nitrate, fluoride, chloride and thiocyanate species sorbed by Dowex 1 have been identified through their spectra. In neutral solutions the Dowex A-1 resin sorbs uranyl ions strongly. Here, too, added salt leads to coordination of the neutral salt anion. Such coordination, dependent on the stability of the various complexes, is observed in the absorption spectrum as a red shift. From this shift, the anions may be classified according to their coordinating power: $NO_3^- \approx SO_4^{2-} < Br^- < Cl^- \approx SCN^-$.

Spectral studies that have so far been made of the chelating resins may still be considered preliminary in nature and further research should contribute importantly to a better understanding of the coordination chemistry of the metal ions in these polymeric chelate systems, their bonding parameters and the kinetics of sorption.

D. Spectra of Ion Exchange Membranes

Little is known concerning site binding and the transfer processes in membranes. The very systematic infrared spectral study of water bands that has been made by Zundel et al. (21,21a) on cation exchange membranes to investigate their hydration properties has been informative in this direction. Discussion of this research, however, is limited to the transition metal ion forms of the cation-exchange membranes employed in their study to complement such emphasis of the material already presented.

The stretching vibration frequency of the hydroxyl groups of the molecules of water bound to the transition ions in salts of polystyrenesulfonic acid, were found to shift according to the location of the cation in the periodic table as shown below:

Ion	Mn^{2+}	Co^{2+}	Ni^{2+}	Cu^{2+}	Zn^{2+}
Band position (cm^{-1})	3406	3381	3373	3314	3317

There is a slight decrease in energy proceeding across the periodic table until Zn^{2+} is reached. The calculated electrostatic cation field value (with Coulomb's formula (21a)) in the direction of the bonds at the hydrogen nuclei of the water molecules is larger in the case of magnesium than it is for any divalent cation of the first transition row. In spite of this, for all these cations except Mn^{2+}, the stretching vibration of the hydroxyl group is shifted to lower energies than in the case of Mg^{2+}. The strength of the hydration bridges by which the water molecules are bound to the anions is thus stronger than would normally be expected from the electrostatic field of these cations. Therefore another interaction must exist between these cations and the water molecules to supplement their polarization by the cation field. This additional interaction strengthens the hydrogen bridge donor properties of the hydroxyl group and shifts the stretching band towards lower energies (smaller wave numbers). This effect must be connected with the crystal field energy made available by the rearrangement of the d electrons by the dipole fields of the water molecules. This occurs for all the cations excepting Mn^{2+} and Fe^{3+}, which do not gain energy by this process. A covalent bond is obtained through the interaction of the d electrons with the lone pairs of the water molecules, thus causing an enhancement of the hydrogen bridge donor property of the hydroxyl groups of the water molecules. The shift of the stretching frequency for the first transition metal cations agrees with the stability sequence found for numerous complexes by Irving and Williams (139): $Mn^{2+} < Co^{2+} < Ni^{2+} < Cu^{2+} > Zn^{2+}$. From this Zundel (21a) concluded that the observed band shift is related to the magnitude of the covalent bonding component in the cation-water bond.

Zundel (21a) also compared the intensity of these stretching frequencies by using membranes hydrated at comparable atmospheric humidities. It was found that the intensities vary as follows: $Mn^{2+} < Co^{2+} < Ni^{2+}$ and Cu^{2+}, $Zn^{2+} < Co^{2+}$, Ni^{2+}. At low humidity the intensity for Zn^{2+} is lower than it is for Cu^{2+}, while at a high degree of humidity the opposite occurs. It follows from these observations that although the molecules of water are more strongly attached in the case of Cu^{2+} and Zn^{2+} than in the cases of Co^{2+} and Ni^{2+} (as seen from the frequency shift), the number of molecules of H_2O bound at the same degree of hydration of the membrane will be smaller for Cu^{2+} and Zn^{2+} than for Co^{2+} and Ni^{2+}. Thus, it is suggested that while six-coordinated octahedral species are formed for Co^{2+} and Ni^{2+}, four-coordinated species, quadratically planar for Cu^{2+} and tetrahedral for Zn^{2+} are formed. Such tetra-coordinated species have not been identified in aqueous solution and they are unlikely to occur in the membrane as well. It is possible that additional coordination to the sulfonate groups occurs through coulombic interaction. Measurements in the visible region of the spectrum are needed to provide the answer to this question.

A membrane study (140) has recently been initiated in the author's laboratory in which the visible and infrared spectra of sorbed complex species are to be compiled, in order to determine the identity of the species sorbed and the type of bonding between species and exchange site. Some preliminary results have been obtained for the cobalt nitrate and chloride systems on an anion-exchange membrane. With the chloride form of the anion-exchange membrane, $CoCl_4^{2-}$ was identified through its spectrum (Fig. 27), even though the predominant ion of the equilibrium solution was $Co(H_2O)_6^{2+}$. The spectrum of the membrane sample, compared with that of the respective solutions, shows a small shift to lower peak energies.

The $Co(NO_3)_4^{2-}$ species was identified only on a vacuum-dried anion-exchange membrane (Fig. 28). When the membrane is left exposed to air the complex is partially decomposed. Through IR measurements it was possible to differentiate between the chloride and nitrate form of the anion-exchange membrane as well as between free NO_3^- ions and coordinated nitrato ions. It is expected that a continuation of this work will contribute to a better understanding of membrane processes.

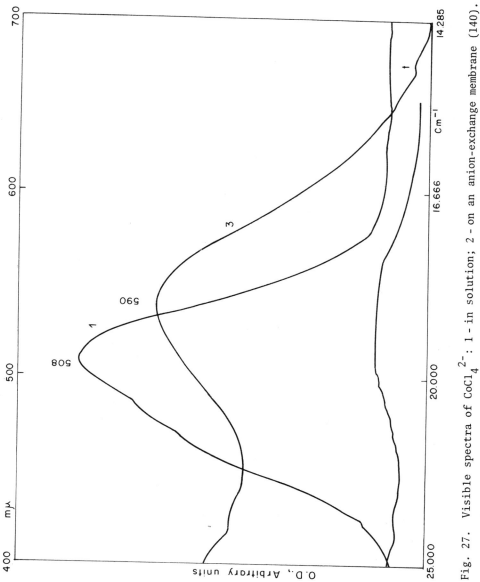

Fig. 27. Visible spectra of $CoCl_4^{2-}$: 1 - in solution; 2 - on an anion-exchange membrane (140).

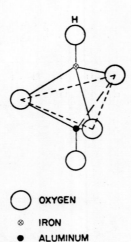

Fig. 28. Visible spectra of Co(NO$_3$)$_2$ in NaNO$_3$ solution: 1 - in solution; 2 - sorbed on a wet anion-exchange membrane; 3 - sorbed on a dry anion-exchange membrane (140).

E. Concluding Remarks

The extent to which spectral techniques have contributed to a) the identification of complex species sorbed on exchangers and to b) the understanding of interactions at ion-exchange sites has been clearly demonstrated.

In general, the species sorbed on a cation or anion exchanger have the same coordination, symmetry and type of bonding properties as those known for other media, i.e. no distortion is observed during sorption. Proof for the sorption of an undistorted complex species has been obtained by comparing the various quantitative parameters evaluated from spectral measurements of the ion exchanger with those calculated for solids or other media. This observation applies for organic cation and anion exchangers, membranes and even zeolites in the hydrated form. It follows from this that the undistorted cationic or anionic species thus sorbed is bound electrostatically to the particular fixed functional unit of the matrix.

SPECTROSCOPIC STUDIES OF ION EXCHANGERS

If in a stepwise reaction cationic, neutral and anionic complex species are formed, it was found that in general, the highest negatively charged species is sorbed even though this species is not always found to exist in solution. Examples of this are $Co(NO_3)_4^{2-}$ (96) and $UO_2(SO_4)_3^{4-}$ (105). A question one is tempted to ask is: why are such unstable species as these sorbed on the exchanger? A reasonable answer to this question may be that:

1) The large fixed ion, which in the case of Dowex 1 is a large cation, i.e. $CH_2N(CH_3)_3^+$, is an important factor. The size of the cation seems to be critical, since all the nitrato-cobalt complexes that have been prepared used large cations for their preparation, e.g. $N(CH_3)_4^+$ and $(C_6H_5)_3AsCH_3^+$ (141);

2) The much smaller solvation shell (i.e. hydration shell) in the resin, compared to the solution, is even more diminished by the presence of a large excess of neutral salt.

The first positively charged species, sorbed on the cation-exchange resin, membrane and hydrated zeolites, is the hexahydrated ion, which, again, has all the spectral characteristics of the species present in aqueous solution. This also implies an electrostatic type of bonding to the fixed anion in the resin. In an NMR study by Creekmore and Reilley (75), the small difference in the hydration number of Na^+ between a solution and the ion exchanger was interpreted as being due to the participation of the sulfonic group of the resin in the coordination of three molecules of water instead of the four that coordinate in solution. The bonding of RSO_3^- to Na^+ is not covalent but a result of coulombic interaction. Similar results have been obtained for transition cations sorbed on dehydrated zeolites, cation-exchange resins and cation-exchange membranes (54-57,88,9-12,21a). These results lead to the same estimate of the situation: There is electrostatic interaction between the cation and the sulfonic group which leads to a somewhat depleted hydration shell. The formation of hydrated cations which are four coordinated are ascribed to Co^{2+} (54-57), Cu^{2+} and Zn^{2+} (21a). These species are not known in solution where an abundance of water is available for a full coordination of six (21a). Further proof for the existence of these ions which has been deduced from IR measurements or Mössbauer spectra measurements (88) on the exchangers has not yet been forthcoming. It is possible that very careful spectral measurements in the visible region of cation exchangers of controlled water content could be employed for this purpose. Changes in coordination number from six to four in transition ions are quite drastic and should be easily detectable. It is possible, however, that although there are only four water molecules, two further coordination sites may be provided by the sulfonic groups electrostatically bound.

V. ADDENDUM

Since preparation of this chapter additional significant research, especially with the zeolites, has been reported. The main objective of this research has been to correlate reactivity with deductions derived from spectroscopic examination of the zeolites containing various cations and adsorbed gaseous molecules. There has, as a consequence, been additional information obtained with regard to the coordination chemistry of the transition metals in zeolites, and the spectral properties defined.

Kellerman et. al. (142) have continued their work on cobalt zeolite complexes with unsaturated hydrocarbons (ethylene propylene and acetylene. Generally, one hydrocarbon molecule is π bonded per cobalt ion which will be in a somewhat distorted tetrahedral site (3 oxygens from the zeolite framework providing the three other ligands). A theoretical analysis which treats the unsaturated hydrocarbon molecule as a low symmetry electrostatic perturbation to a tetrahedral cobalt ion provides the interpretation of the band splitting of the electronic spectra. In another paper by the same groups (143) it has been found that divalent chromium sorbed on an anhydrous zeolite is a reversible oxygen binder. Here too, as in the cobalt case, one molecule of oxygen is π bonded to the chromous ion by accepting an electron, and thus a trivalent chromium ion is obtained. The spectral and magnetic properties are also discussed.

Extensive study of this subject has been carried out by Lunsford and his group (144-147) mainly by IR and ESR spectra; reference (146) reviews part of this work. The transition metals studied were cobalt, ruthenium, osmium, copper, silver and iron, while the ligands were nitrosyl, dinitrogen, ammines, hydrogen and oxygen. From these studies it may be concluded as before that in some cases the complexes formed are the same as in conventional solvents while in others the zeolite may also function as a ligand in defining the coordination site of the ion.

Infrared studies, for the primary purpose of elucidating the framework structure under different conditions (various ions, temperature and pressure), has been extended by other researchers (148-150). Mörke et al. (151) have studied the coordination sites of copper in the NaY zeolite at various temperatures and copper concentrations by electron spin resonance. The sites were identified by their accessibility to reduction by H_2 and CO. The formation of clusters was found to occur in the supercage, being favored by an increased copper concentration or by dehydration. From the numbers and values of the g parameters the ligand invironment was determined in each temperature range.

Boudart et al.(152) have studied the reversible oxidation of ferrous ions in Y zeolites by far infrared measurements. The uptake of oxygen is one to two iron ions implying the formation of an iron bridge Fe-O-Fe with Fe^{3+} occupying sites S_1' of two adjacent hexagonal prisms. Progress has also been made in the study of the Raman spectrum of zeolites. These measurements have been difficult to make because of the fluorescence of the zeolite samples. This fluorescence has been attributed (153) to the presence of traces of unsaturated hydrocarbons. By igniting the zeolite samples in oxygen at 500°C it has been possible to remove these traces to reduce appreciably the fluorescence (thereby rendering feasible the Raman measurements). The adsorption of pyridine was studied on the NaY zeolite (153) where Na ions were replaced by Li, K, Co, Cu, Ce or H ions. The position of the Raman band suggests that pyridine adsorption is essentially on the cationic site.

Significant new research has also been reported for organic ion exchangers (154-155). Waki et al.(154) have studied the stability of the complexes of copper tetraamine ions with linear phosphate ions. They have determined the stability constants by cation-exchange distribution methods. These results have been correlated with the spectra of these complexes in solution and sorbed on an anion exchange resin. The same authors (155) have applied the measurement of the spectra of ionic species sorbed on resins to the development of colorimetric micro-determinations of Cr, Fe, Co and Cu. The sensitivity obtained by these methods is about ten times that obtained with conventional solution colorimetry.

ESR studies of manganese sorbed on cation exchangers of different water content have been made by Goldammer et al. (156). At high water concentration the measurements indicate that the hydrated manganese molecule is very mobile in the polymer matrix, while at lower water content (less than three water molecules per manganese ion) there is an interaction with the sulfonic group.

In this laboratory (157) the sorption of cobalt ions on a sulfonic resin of various degrees of crosslinking has been studied by means of spectral measurements mainly in the visible and in the infrared region. The purpose of this study was to determine the type of species of cobalt sorbed on this resin as function of the degree of crosslinking, as well as its type of binding. In some of the experiments a fixed amount of neutral chloride salt was added in order to examine the influence of the chloride ion on the species formed. From these measurements it can be concluded that after a certain drying interval tetrahedral cobalt species are formed at each crosslinking, the most rapid formation of these species being observed on the lowest crosslinked resin samples. According to its spectrum, the tetrahedral species formed on low and high crosslinked species are not the same, however. If the spectra are interpreted according to the spectrochemical series, it may be concluded that a tetrahedral tetra-aquo $Co(H_2O)_4^{2+}$ or $Co(H_2O)_5Cl^+$ (when chloride salt is added) is sorbed on a low crosslinked resin while on the high (12%) crosslinked resin the cobalt interacts strongly with the sulfonate group. These results agree with the IR measurements which show shifts in the sulfonic band frequencies. The NMR measurements made earlier on alkali and alkaline earth ions sorbed on cation exchange resins (75-80) also imply an interaction with the sulfonic group which replaces one molecule of water; however, electrostatic interaction was assumed. This apparently is not the case for the cobalt ion.

Spectral studies of cation and anion exchange membranes to study the water sorption as well as the sorption of various complex species of transition metals, are being continued in this laboratory (158). All the regions of the spectrum (ultraviolet, visible, near infrared, infrared and far infrared) have been used with good success on all the membrane samples. In some cases the tetrahalides of Co and Ni were sorbed while in some of the cases interaction with the sulfonamide site (for the cobalt ions) has been assumed.

REFERENCES

1. E. Rutner, J. Phys. Chem., 65, 1027 (1961).
2. K. Klier and M. Ralek, J. Phys. Chem. Solids, 29, 951 (1968).
3. G. Kortüm, W. Braun and G. Herzog, Angew. Chem., 75, 653 (1963).
4. B. Wichterlova, P. Jiru and A. Curinova, Z. fur Phys. Chem., 88, 180 (1974).
5. E. Garbowski, Y. Kodratoff, M. V. Mathieu and B. Imelik, J. Chim. Phys., 69, 1386 (1972).
6. J.S. Coleman, J. Inorg. Nucl. Chem., 28, 2371 (1966).
7. J.L. Ryan, Inorg. Chem., 2, 348 (1963).
8. J.L. Ryan, J. Phys. Chem., 65, 1099 (1961)
9. T. Nortia and S. Laitinen, Suomen Kem., B41, 136 (1968).

10. S. Laitinen and T. Nortia, Suomen Kem., ibid., B41, 253 (1968).
11. S. Laitinen and T. Nortia, ibid., B43, 128 (1970).
12. S. Laitinen and T. Nortia, ibid., A44, 79 (1971).
13. H. Waki, S. Takahasi and S. Ohashi, J. Inorg. Nucl. Chem., 35, 1259 (1973).
14. C. Heitner-Wirguin and R. Cohen, J. Phys. Chem., 71, 2556 (1967).
15. N. Ben-Zwi and C. Heitner-Wirguin, Israel J. Chem., 10, 885 (1972).
16. O. Lahodny-Sarc and J.L. White, J. Phys. Chem., 75, 2408 (1971).
17. A.A. Kubasov, K.V. Topchieva and A.N. Ratov, Russ. J. Phys. Chem., 47, 1023 (1973).
18. J.W. Ward, J. Phys. Chem., 74, 3021 (1970).
19. J.W. Ward, J. Catalysis, 20, 1489 (1971).
20. R.A. Schoonheydt and J.B. Uytterhoeven, J. Catalysis, 19, 55 (1970).
21. G. Zündel, H. Noller and G.M. Schwab, Z. Naturforsch, 16B, 716 (1961).
21a. G. Zündel, Hydration and Intermolecular Interaction, Acad. Press, New York, 1969.
22. P. Dobud, D. Sutton and D.J. Tuck, Canad. J. Chem., 48, 2290 (1970).
23. G.E. Boyd, private communication (1971).
24. B.D. McNicol, G.T. Pott and K.R. Loos, J. Phys. Chem., 76, 3388 (1972).
25. C.L. Angell, J. Phys. Chem., 77, 222 (1973).
26. W.N. Delgass, R.L. Garten and M. Boudart, J. Chem. Phys., 50, 4603 (1969).
27. W.N. Delgass, R.L. Garten and M. Boudart, J. Phys. Chem., 73, 2970 (1969).
28. J.V. Smith, Mineralogical Soc. Am. Spec. Paper No. 1, 281 (1963).
29. K.R. Fischer, W.M. Meier, Fortschr. Mineral, 42, 50 (1965).
30. W.M. Meier, D.H. Olson, Advan. Chem. Ser., 101, 155 (1971).
31. R.G. Milkey, Am. Mineralogist, 45, 990 (1960).
32. A.V. Kiselev and V.J. Lygin, in Infrared Spectra of Adsorbed Species, L.H. Little, Ed., pp. 361-7, Academic Press, London, 1967.
33. A.C. Wright, J.P. Rupert and W.T. Granquist, Am. Mineralogist, 53, 1293 (1968).

34. S.P. Zhdanov, A.V. Kiselev, V.J. Lygin and I.I. Titova, Russ. J. Phys. Chem., 38, 1299 (1964).
35. V.H. Dutz, Ber. Deut. Keram. Ges., 46, 75 (1969).
36. E.M. Flanigen, H. Khatami and H.A. Szymanski, Adv. in Chem. Series, 101, 201 (1971).
37. K. Nakamoto, Infrared Spectra of Inorganic and Coordination Compounds, pp. 103-107, Wiley, New York, 1963.
38. B.D. Saksena, Trans. Faraday Soc., 57, 252 (1961).
39. H.P. Gregor, J. Am. Chem. Soc., 70, 1293 (1948).
40. H.P. Gregor, ibid., 73, 642 (1951).
41. E. Glueckauf, Proc. Roy. Soc. London, A214, 207 (1952).
42. S. Bukata and J.A. Marinsky, J. Phys. Chem., 68, 994 (1964).
43. H.A. Szymanski, D.N. Stamires and G.R. Linch, J. Opt. Soc. Am., 50, 1323 (1960).
44. H.S. Sherry, Ion Exchange, Vol. 2, p. 89, ed. J.A. Marinsky, Dekker, 1969.
45. G.E. Boyd, A.W. Adamson and L.S. Myers, J. Am. Chem. Soc., 69, 2836 (1947).
46. J.W. Ward, Adv. in Chem. Series, 101, 380 (1971).
47. L. Bertsch and H.W. Habgood, J. Phys. Chem., 67, 1621 (1963).
48. J.W. Ward, J. Phys. Chem., 72, 4211 (1968).
49. H.A. Resing and J.K. Thompson, Adv. in Chem. Series, 101, 473 (1971).
50. R.J. Faber and M.T. Rogers, J. Am. Chem. Soc., 81, 1849 (1959).
51. T.I. Barry and L.A. Lay, J. Phys. Chem. Solids, 29, 1395 (1968).
52. L.G. Dzhashiashvili, N.N. Tikhomirova and G.V. Tsitsishvili, Russ. J. Struct. Chem., 10, 364 (1969).
53. J.A. Morice and L.V.C. Rees, Trans. Faraday Soc., 64, 1388 (1968).
54. T.A. Egerton, A. Hogan, F.S. Stone and J.C. Vickerman, J. Chem. Soc., Faraday Trans. I, 68, 723 (1972).
55. P. Gallezot and B. Imelik, J. Chim. Phys., 71, 155 (1974).
56. K. Klier, J. Am. Chem. Soc., 91, 5392 (1969).
57. K. Klier, Adv. in Chem. Series, 101, 48 (1971).
58. R. Polak and V. Cerny, J. Phys. Chem. Solids, 29, 945 (1968).
59. R. Polak and K. Klier, J. Phys. Chem. Solids, 30, 2231 (1969).
60. M. Guilleux and J. Tempere, Comp. Rend. Acad. Sci. Paris, C272, 2105 (1971).

61. J.D. Mikheikin, V.A. Svets and V.B. Kazanskii, Kinetika i Kataliz, 11, 747 (1970).
62. N.N. Tikhomirova and J.V. Nikolaeva, Russ. J. Struct. Chem., 10, 457 (1968).
63. J. Turkevich, Y. Ono and J. Soria, J. Catalysis, 25, 44 (1972).
64. W.G. Maksimov, V.F. Anufrienki, K.G. Jone, N.E. Shestakova, Russ. J. Struct. Chem., 13, 953 (1972).
65. J.C. Vedrine, E.G. Derouane and Y. Ben Taarit, J. Phys. Chem., 78, 531 (1974).
66. Y. Yamada, Bull. Chem. Soc. Japan, 45, 60 (1972).
67. Y. Yamada, ibid., 45, 64 (1972).
68. F.A. Cotton and G. Wilkinson, Advanced Inorganic Chemistry, Interscience Publ., 3rd Ed., 1972.
69. L. Sacconi, Pure and Appl. Chem., 17, 97 (1968).
70. M. Ciampolini, Structure and Bonding, 6, 52 (1969).
71. L. Sacconi, J. Chem. Soc. (A), 248 (1970).
72. P.L. Orioli, Coord. Chem. Rev., 6, 285 (1971).
73. G. Zundel and A. Murr, J. Chim. Phys., 66, 246 (1969).
74. R. Cohen and J. Peretz, Israel J. Chem., 7, 667 (1969).
75. R.W. Creekmore and C.N. Reilley, Anal. Chem., 42, 570 (1970).
76. R.W. Creekmore and C.N. Reilley, ibid., 42, 725 (1970).
77. T.E. Gough, H.D. Sharma and N. Subramanian, Can. J. Chem., 48, 917 (1970).
78. H.D. Sharma and N. Subramanian, ibid, 49, 457 (1971).
79. H.D. Sharma and N. Subramanian, ibid., 49, 3948 (1971).
80. L.S. Frankel, J. Phys. Chem., 75, 1211 (1971).
81. A. Narebska and K. Erdman, Rocz Chemii, 47, 1039 (1973).
82. P. Weiner and D.G. Howery, Can. J. Chem., 49, 2913 (1971).
83. T.A. Karpukhina, E.D. Kiseleva, K.V. Chmutov and M.P. Glazunov, Russ. J. Phys. Chem., 44, 557 (1970).
84. J.A. Marinsky and E. Högfeldt, Chemica Scripta, 9, (1976).
85. Y. Marcus and A.S. Kertes, Ion Exchange and Solvent Extraction of Metal Complexes, Wiley, 1969, p. 233.
85a. Y. Marcus in Ion Exchange, Vol. I, ed. J.A. Marinsky, 1966, Marcel Dekker, Inc., p. 101.
86. S. Lindenbaum and G.E. Boyd, J. Phys. Chem., 67, 1238 (1963).
87. Y. Takashima, Y. Maeda and S. Umemoto, Bull. Chem. Soc. Japan, 42, 1760 (1969).

88. A. Johansson, J. Inorg. Nucl. Chem., 31, 3273 (1969).
89. T. Nortia and E. Kontas, Suomen Kem., 44, 406 (1971).
90. G.E. Stein and J.A. Marinsky, J. Inorg. Nucl. Chem., 37, 2421(1975).
91. C. Furlani and G. Morpurgo, Theor. Chim. Acta, 1, 102 (1963).
92. J. Ferguson, J. Chem. Phys., 40, 3406 (1964).
93. S.E. Manahan and R.T. Iwamoto, Inorg. Chem., 4, 1409 (1965).
94. C. Heitner-Wirguin and N. Ben-Zwi, Inorg. Chim. Acta, 4, 517 (1970).
95. C. Heitner-Wirguin and N. Ben-Zwi, ibid., 4, 554 (1970).
96. C. Heitner-Wirguin and N. Ben-Zwi, J. Inorg. Nucl. Chem., 33, 1493 (1971).
96a. N. Ben-Zwi, Ph.D. Thesis, Hebrew University, Jerusalem (1971).
97. C. Heitner-Wirguin and N. Ben-Zwi, Inorg. Chim. Acta, 6, 93 (1972).
98. C.C. Addison and B.M. Gatehouse, J. Chem. Soc., 613 (1960).
99. H.L. Schläfer and G. Gliemann, Basic Principles of Ligand Field Theory, Wiley Interscience, 1969.
100. A. Turco and C. Pecile, Nature, 191, 66 (1961).
101. A. Turco, C. Pecile and N. Nicolini, J. Chem. Soc., 3008 (1962).
102. P.C.H. Mitchell and R.J.P. Williams, J. Chem. Soc., 1912 (1960).
103. R. Tsuchida, Bull. Chem. Soc. Japan, 13, 388, 436, 471 (1938).
104. C. Schäffer and C.K. Jörgensen, J. Inorg. Nucl. Chem., 8, 143 (1958).
105. C. Heitner-Wirguin and M. Gantz, J. Inorg. Nucl. Chem., 35, 3341 (1973).
106. M. Gantz, Ph.D. Thesis, Hebrew University, Jerusalem (1972).
107. C. Heitner-Wirguin and M. Gantz, Israel J. Chem., 12, 723 (1974).
108. S.P. McGlynn and J.K. Smith, J. Mol. Spectrosc., 6, 164 (1961).
109. J.T. Bell and R.E. Biggers, J. Mol. Spectrosc., 18, 247 (1965).
110. C. Heitner-Wirguin, Lecture presented at NATO Advanced Study Institute on Charged and Reactive Polymers, Forges, 1973, D. Reidel Publ. Co.
111. C. Görller-Walrand and S. DeJaegere, Spectrochim. Acta, 28A, 257 (1972).
112. C. Heitner-Wirguin and G. Markovitz, J. Phys. Chem., 67, 2263 (1963).
113. A. Schwarz, J.A. Marinsky and K.S. Spiegler, J. Phys. Chem., 68, 918 (1964).

114. C. Heitner-Wirguin and V. Urbach, J. Phys. Chem., 69, 3400 (1965).
115. C. Heitner-Wirguin and J. Kendler, J. Inorg. Nucl. Chem., 33, 3119 (1971).
116. E.A. Chuveleva, N.K. Yufryakova, P.P. Nazarov and K.V. Chmutov, Russ. J. Phys. Chem., 44, 1125 (1970).
117. E.A. Chuveleva, N.K. Yufryakova, P.P. Nazarov and K.V. Chmutov, ibid., 46, 700 (1972).
118. N.K. Yufryakova, E.A. Chuveleva, P.P. Nazarov and K.V. Chmutov, ibid., 46, 702, 705 (1972).
119. R. Cohen and C. Heitner-Wirguin, Inorg. Chim. Acta, 3, 647 (1969).
119a. R. Cohen, Israel J. Chem., 9, 499 (1971).
120. K. Umezawa and T. Yamabe, Bull. Chem. Soc. Japan, 45, 56 (1972).
121. M. Mandel and J.C. Leyte, J. Polymer Sci., A2, 2883 (1964).
122. M. Mandel and J.C. Leyte, J. Polymer Sci., A2, 3771 (1964).
123. J.C. Leyte, L.H. Zuiderweg and H.J. Vledder, Spectrochim. Acta, 23A, 1397 (1967).
124. J.C. Leyte, L.H. Zuiderweg and M. van Reisen, J. Phys. Chem., 72, 1172 (1968).
125. J.C. Leyte, Polyelectrolytes (ed. E. Selegny), D. Reidel Publ. Co., Holland, Vol. 1, pp. 339-346.
126. See ref. 125, J.C. Leyte, L.H. Zuiderweg and J.J. van den Klink, pp. 383-389.
127. J.A. Marinsky, N. Imai and M.C. Lim, Israel J. Chem., 11, 601 (1973).
128. C. Travers and J.A. Marinsky, J. Polymer Sci. Symposium, 47, 285 (1974).
129. W.M. Anspach and J.A. Marinsky, J. Phys. Chem., 79, 433 (1975).
130. J.A. Marinsky and W.M. Anspach, J. Phys. Chem., 79, 439 (1975).
131. P. Spegt and G. Weill, Compt. Rend. Acad. Sci. Paris, 274C, 587 (1972).
132. See ref. 125, G. Weill and P. Spegt, pp. 371-382.
133. V.J. Paramonova, G.P. Nikitina and G.P. Akonov, Radiokhimiya, 10, 646 (1968).
134. K.M. Saldadze, V.D. Kopylova, T.V. Mekvabishvili and R.J. Machkhoshvili, Russ. J. Phys. Chem., 45, 672 (1971).
135. E.A. Chuveleva, N.K. Yufryakova, P.P. Nazarov and K.V. Chmutov, ibid., 46, 51 (1972).
136. C. Heitner-Wirguin and N. Ben-Zwi, Israel J. Chem., 8, 913 (1970).

SPECTROSCOPIC STUDIES OF ION EXCHANGERS

137. J. Krasner and J.A. Marinsky, J. Phys. Chem., 67, 2559 (1963);
137a. C. Eger, W.M. Anspach and J.A. Marinsky, J. Inorg. Nucl. Chem., 30, 1899 (1968);
137b. C. Eger, J.A. Marinsky and W.M. Anspach, ibid., 30, 1911 (1968).
138. G. Schmuckler, Talanta, 10, 745 (1963);
138a. H. Loewenschuss and G. Schmuckler, ibid, 11, 1399 (1964).
138b. G. Schmuckler, ibid, 12, 281 (1965).
139. H. Irving and R.J.R. Williams, Nature, 162, 746 (1948).
140. C. Heitner-Wirguin and D. Hall, unpublished results.
141. F.A. Cotton and T.G. Dunne, J. Am. Chem. Soc., 84, 2013 (1962).
142. K. Klier, R. Kellerman and P.J. Hutta, J. Chem. Phys., 61, 4224 (1974).
143. R. Kellerman, P.J. Hutta and K. Klier, J. Amer. Chem. Soc., 96, 5946 (1974).
144. K.A. Windhorst and J.H. Lunsford, J. Amer. Chem. Soc., 97 1407 (1975).
145. K.R. Laing, R.L. Leubner and J.H. Lunsford, Inorg. Chem., 14, 1400 (1975).
146. J.H. Lunsford, Catal. Rev. - Sci. Eng., 12, 137 (1975).
147. R.G. Herman, J.H. Lunsford, H. Beyer, P.A. Jacobs and J.B. Uytterhoeven, J. Phys. Chem., 79, 2388 (1975).
148. E. Gallei and D. Eisenbach, J. of Catalysis, 37, 474 (1975).
149. J. Scherzer, J.L. Bass, F.D. Hunter, J. Phys. Chem., 79, 1194 (1975).
150. J. Scherzer and J.L. Bass, J. Phys. Chem., 79, 1200 (1975).
151. W. Mörke, F. Vogt and H. Bremer, Z. Anorg. Allg. Chem., 422, 273 (1976).
152. R.A. Dalla Betta, R.L. Garten and M. Boudart, J. of Catalysis, 41, 40 (1976).
153. T.A. Egerton, A.H. Hardin and W. Sheppard, Can. J. Chem., 36, 1337 (1974).
154. H. Waki, K. Yoshimura and S. Ohashi, J. Inorg. Nucl. Chem., 36, 1337 (1974).
155. K. Yoshimura, H. Waki and S. Ohashi, Talanta, 23, 449 (1976).
156. E. von Goldammer, A. Müller and B.E. Conway, Ber. Bensenges Physik. Chem., 78, 35 (1974).
157. C. Heitner-Wirguin, to be published.
158. D. Hall, Ph.D. Thesis, Hebrew University, Jerusalem Israel (1976).

Chapter 4

ION-EXCHANGE MATERIALS IN NATURAL WATER SYSTEMS

Michael M. Reddy
Division of Laboratories and Research
New York State Department of Health
Albany, New York

I. INTRODUCTION ... 166
II. ION-EXCHANGE REACTIONS IN NATURAL WATERS ... 168
   A. Ion-Exchange Materials ... 169
      1. Clay minerals ... 169
      2. Other natural components ... 171
   B. Ion-Exchange in Sediments ... 173
III. RELATIONSHIP BETWEEN CLAY MINERAL STRUCTURE AND ION-EXCHANGE PROPERTIES ... 175
   A. Structure - Ion-Exchange Relationship for Kaolinite ... 176
   B. Structure - Ion-Exchange Relationship for More Complex Clays ... 176
IV. ION-EXCHANGE EQUILIBRIA AND SELECTIVITY EXPRESSIONS FOR COMPONENTS OF NATURAL EXCHANGE SYSTEMS ... 179
   A. Affinity Series for Ions on Natural Ion-Exchange Materials ... 180
   B. Experimental Measurements Used to Characterize Ion-Exchange Equilibrium and Selectivity for Natural Materials ... 181
   C. Treatment of Ion-Exchange Equilibria and Selectivity ... 182
   D. Development of Expressions for Presenting Experimental Ion-Exchange Selectivity Data ... 184
   E. Thermodynamic Treatment of Ion-Exchange Selectivity as a Function of Solution Normality ... 188
   F. Thermodynamic Treatment of Incomplete Exchange ... 192
V. EXPERIMENTAL MEASUREMENTS OF ION-EXCHANGE SELECTIVITY FOR COMPONENTS OF NATURAL WATER SYSTEMS ... 193
   A. Selectivity Measurements for Pure Reference Clays ... 193
   B. Selectivity Measurements for Purified Natural Materials ... 198

C. Selectivity Measurements for Natural Materials 202
D. Selectivity Measurements for Sediments 210
VI. SUMMARY 215
REFERENCES 215

I. INTRODUCTION

The composition of natural water systems is significantly influenced by the presence of ion-exchange materials. Ion-adsorption and-exchange processes involving natural waters and their sediments help determine the geochemical distribution of elements released by weathering of the earth's crust. Furthermore, the natural chemical cycles of some ionic pollutants undoubtedly involve ion-exchange reactions, and knowledge of these processes in natural systems is necessary for evaluating the fate of pollutants in the environment. Gibbs [1] has recently shown that for two large river systems the transport mechanism for several trace metals involves four chemically different phases associated with the suspended sediment: an ion-exchange phase, a metallic oxide phase, an organic phase, and a crystalline matrix phase.

The principles and practice of ion exchange are described in several monographs. A detailed, comprehensive analysis of ion exchange has been prepared by Helfferich [2]. Amphlett [3] surveyed inorganic ion exchangers, and Kelly [4] reviewed much of the earlier literature with emphasis on ion exchange in soils. In an investigation of radioactive waste disposal, Robinson [5] reviewed ion-exchange minerals and their interaction with ionic substances in solution.

For natural water systems, however, a detailed description of the many ion-exchange processes is only partially possible at the present time because of the many laboratory and field investigations that still need to be performed. Experimental data have been analyzed using a number of mathematical approaches, making the

ION-EXCHANGE MATERIALS IN NATURAL WATER SYSTEMS

studies difficult to compare. While results of several investigators could be compared using an exact thermodynamic formulation such as that proposed by Gaines and Thomas [6,7], only a few comprehensive studies of ion-exchange substances have used such a rigorous thermodynamic approach.

Some recent laboratory and field studies have addressed specific areas of interaction between solids and natural water, such as the interaction of phosphate and other nutrients with soils and sediments and the accumulation of trace metal ions by iron and manganese oxides. A common defect of these studies is the absence of a clear definition of the interaction. Too often the description of the solid phase is unclear, and thermodynamic equilibria and reversibility have not always been verified. A complete description of ion exchange in natural waters cannot be accomplished without considering these important areas.

Ion-exchange reactions in natural systems differ in some ways from reactions of pure materials under laboratory conditions. In natural systems ion-exchange materials are not pure, well-defined substances. They may be mixtures of crystalline or amorphous materials and often contain mixed-layer clays, mixtures of clays, metallic oxides, and organic substances. Despite the apparent complexity of natural ion exchangers, however, a specific substance or phase usually dominates the ion-exchange behavior for a given ion [8].

The most substantial amount of data about natural ion-exchange materials has been obtained for clay minerals, in which ion-exchange reactions often proceed reversibly and stoichiometrically. However, certain minerals exhibit irreversible ion-exchange behavior, while others show nonstoichiometric exchange in acid solution by releasing aluminum- and silicon-containing ions. The reactions of some natural ion-exchange materials differ from an ideal ion exchanger but to a small enough degree that they are still recognizable as ion-exchange reactions.

Biological processes in aquatic systems intervene in ionic redistribution in numerous ways, for example, by altering solution pH or Eh, or affecting the selective uptake of ions. A detailed description of these processes is outside the scope of the present chapter, although they are obviously important in many areas of natural water chemistry.

In this chapter we will examine the properties of natural ion-exchange materials with emphasis on clay minerals and evaluate the suitability of physicochemical descriptions of ion exchange for materials in natural waters.

II. ION-EXCHANGE REACTIONS IN NATURAL WATERS

Ion-exchange reactions in natural waters include redistribution of ions between soil and soil-pore water, between rock and rock-pore water, and between water and suspended sediment. Reactions taking place during sediment accumulation, halmyrolysis, and diagenesis also appear to involve ion exchange. Halmyrolysis refers to sediment reactions that occur before sediment is buried [9]; diagenesis refers to all processes that change fresh sediment into rock at surface temperature and pressure [10]. An integral part of sediment diagenesis is gravitational compaction, which causes a decrease in sediment pore volume and release of pore fluids. The importance of ion-exchange and adsorption processes in sediment diagenesis has been pointed out by Berner [11].

Ion-exchange reactions affect the composition of rock- and soil-pore waters. As Hem [12] has noted, sediments deposited in a marine environment and later flushed with fresh water have mostly sodium ions in the exchange sites and exchange them for calcium and magnesium ions in the groundwater. A natural softening effect is the result and is common in deep granular aquifers.

Ion-exchange reactions in natural waters are influenced by (a) the chemical state of the exchanging ion and (b) the physical

and chemical state of the ion-exchange material, as follows:
(a) The chemical state of the exchanging ion can be determined from the total concentration of dissolved metal, the estimated extent of ion association, and activity coefficients. Lee [13] has stated that natural organic materials which lead to ion association in solution affect the distribution of heavy metals associated with natural ion-exchange materials. Ionic interactions in natural waters can be characterized by a calculation scheme such as that proposed by Morel and Morgan [14]. (b) The physical state of a natural ion-exchange material refers principally to its state of aggregation. The chemical state includes the bonding and interactions in the solid material. For most natural ion-exchange substances factors such as temperature, ionic strength, and solution composition partially regulate the chemical and in some instances the physical states.

A. Ion-Exchange Materials

1. Clay Minerals

Ion-exchange reactions involving ionic transport and accumulation are usually associated with a specific phase or material such as a suspended sediment or soil. Clay minerals (hydrous aluminosilicate compounds) are the most abundant minerals in sedimentary rock and frequently occupy a central role in the environmental redistribution of metal ions. They occur in the earth's crust in the order: illite > montmorillonite > mixed-layer illite-montmorillonite > chlorite > mixed-layer chlorite-montmorillonite > kaolinite [15].

The ion-exchange characteristics of clays are affected by many factors including parent material (geology), age of weathering surface, climatic factors, degree of weathering, and completeness of physical dispersion [16]. Clay minerals can be formed by a number of different mechanisms [5], including leaching of certain elements from older minerals, complete replacement of a mineral, devitrification of volcanic glasses, crystallization of colloidal material, direct precipitation, and reactions between older minerals and solutions.

Hendricks [17] was one of the first to propose a useful model of clay ion exchange. He pointed out that an excess of charge on the surface of a clay mineral would require external ions to balance the charge. Frederickson [18] extended this model to weathering processes. He suggested a mechanism of mineral weathering in which hydrogen ions of an aqueous solution surrounding a mineral such as albite are exchanged for replaceable ions (sodium in this case). This substitution causes the crystal to expand, which in turn increases the reactivity of the mineral and thus hastens the collapse of the original mineral matrix.

Recent investigations of mineral weathering and its relation to clay mineral formation have been summarized by Stumm and Morgan [19], who consider incongruent dissolution of aluminosilicates to be one of the most important weathering reactions. It involves transformation of a mineral to a secondary mineral (reaction product) with a corresponding increase in the alkalinity of the dissolved phase. For example, the main alteration product of feldspar weathering is kaolinite, with montmorillonite and micas as possible intermediates. Clays transported by rivers and streams are produced from soils of the drainage basin and from parent rocks by weathering. Millot [20] has examined large-scale mineral weathering and transport processes in the environment and has summarized the two principal weathering mechanisms. The first is predominantly physical in character, producing nearly intact silicates with their lattice structure and major and minor element composition unchanged from that of the parent rock. The second is a chemical mechanism in which hydrolysis of the silicate leads to destruction of the original matrix with release of the constitutent ions of the mineral to solution.

There are substantial difficulties in following clay formation reactions in the laboratory at ambient temperature. Sluggish

ION-EXCHANGE MATERIALS IN NATURAL WATER SYSTEMS

reaction kinetics and the appearance of amorphous or poorly crystalline reaction products complicate clay formation. Hem et al. [21] have recently examined the products of the chemical interaction of aluminum with aqueous silica at low temperature. Electron micrographs of the resulting aluminosilicates indicate some crystallinity, but no x-ray diffraction pattern could be obtained even in material aged four years. Harder [22,23] examined the synthesis of clay minerals at surface temperatures and found that the formation of illite minerals in the environment is related to adsorption of potassium, magnesium, and iron on amorphous (determined by x-ray diffraction analysis) hydroxide materials. A detailed review and analysis of the chemistry of clay systems with particular emphasis on colloidal properties has been presented by van Olphen [24].

2. Other Natural Components

In addition to clay minerals, other components of sediments such as metallic oxides and organic substances have been implicated in ion-exchange reactions.

Hydrous oxides of iron and manganese have been considered important regulators of certain trace elements in natural waters. Lee [13] finds strong evidence that the commonly encountered hydrous metal oxides of iron and manganese are important sinks for heavy metals and that they provide a transport medium for trace metals in nature. This supports Jenne's suggestion [25] that these oxides control the concentrations of several trace metals in soils and water. Anderson and Jenne [26] revised the analytical methodology used to describe the association of heavy metals with metal oxides in natural ion-exchange materials.

A quantitative assessment of ionic uptake is unavailable for a variety of natural water conditions because of the complexities of the hydrous oxide system. Morgan and Stumm [27] have documented the formation of a poorly ordered nonstoichiometric manganese oxide from solution. A recent investigation by Murray [28] deals with the surface chemistry of hydrous manganese dioxide; his results suggest that a high surface charge on this compound relative to silica (which has a similar pH [ZPC]) can explain trace-metal enrichment of manganese-rich sediments. These publications also indicate some of the problems encountered in preparing metallic oxides and characterizing their interactions in solution. A quantitative description of metallic oxide ion-exchange reactions will probably be unavailable for some time.

Organic materials in natural water systems may contribute to observed ion-exchange behavior. Their physical state (dissolved, colloidal, or solid) and their chemical nature as determined by their origin and potential for transformation are important. A subsequent chapter in the next volume of this series will examine the role of organic ion-exchange materials in natural water systems.

Malcolm and Kennedy [16] have estimated the fraction of organic material in ion-exchange reactions in Mattole River sediments. They assume that it contains 50% carbon and has an ion-exchange capacity of 100 meq/100 g and calculate that it contributes 15% of the cation-exchange capacity. Gibbs [1] has examined ion-exchange reactions in trace metal transport by suspended sediment in the Amazon and Yukon Rivers. His results indicate that no single suspended sediment phase transports all of the metal ions studied.

Dawson and Duursma [29] conducted laboratory experiments to study the uptake of radionuclides by sediments and phytoplankton. The uptake was a function of the relative concentration of each material and the different affinities of the sediment and phytoplankton for the radionuclides. Several ions, including cadmium and silver, showed higher affinities for the phytoplankton, while manganese, cobalt, and cesium favored the sediment. These data and the results of other studies show that uptake of certain trace element ions by organisms in natural waters can be substantial.

Laboratory investigation of natural ion-exchange materials is hindered by the subtle differences between natural substances and the purified materials that are often utilized. Transfer of a natural ion-exchange material to the laboratory involves sampling and storage, which may alter the physical and chemical properties of the material. If a hydrated exchanger is dried, there may be irreversible dehydration of hydrous oxides, oxidation of reduced metal ions and organics, and precipitation of sparingly soluble salts and trace metals [8]. While conditions which lead to uncertainties in the experimental data cannot be completely eliminated, they can be minimized by appropriate experimental procedures.

B. Ion-Exchange in Sediments

Sediments in natural water systems act both as a sink for particulate material settling out of the water and as a source of dissolved substances arising from diagenesis and compaction of the deposited material. They are an integral part of the cycle of elements in the environment and reflect the nature of the overlying water at the time of deposition. Ion-exchange processes are involved in the interaction between sediment and water in natural systems [11].

Hydrologic conditions can affect ion-exchange reactions in natural systems. Stream sediments, for example, are suspended and entrained during periods of high flow, and have a more direct effect on the stream water chemistry than bed sediments. Indeed, at a high suspended sediment content an appreciable fraction of the transported cations are carried as exchange ions with the suspended sediment [30].

Kamp-Nielsen [31] examined three factors regulating the exchange rates of ions across the sediment-water interface: sediment oxidation state, microbiological activity, and pH. He found that adsorption and the concentration gradient of dissolved species were influenced by a pH-dependent dissolution and precipitation of slightly soluble iron and calcium phosphates. Early diagenesis in a reduced fjord sediment has been studied by Presley et al. [32], whose results show that trace-element enrichment of the interstitial water relative to the overlying sea water is related either to organic complexing of the metals or to equilibration with unidentified mineral phases.

Malcolm and Kennedy [16,30] and Jenne and Wahlberg [8] have applied ion-exchange principles to sediment investigations. Sand- and gravel-size sediments show remarkably high cation-exchange capacities (7-16 meq/100 g [16]). Jenne and Wahlberg followed the process by which ^{90}Sr, ^{60}Co, and ^{137}Cs are bound to river sediments. Strontium was found to be associated with carbonate minerals, cobalt with manganese and iron oxides, and cesium with certain clay minerals. The mechanism of cesium uptake involved ion exchange on the clay mineral.

The composition of the interstitial solution of sediments must be known in order to understand ion-exchange reactions in sediments. Siever et al. [33] have discussed the role of interstitial water in modern sediments. Magnesium ion is slightly depleted from the overlying water as a result of uptake by chlorite, while potassium ion is enriched in these solutions from hydrolysis of finely divided detrital K-feldspars. Manheim [34] has examined the current state of interstitial water analysis in the light of ion-exchange reactions between the interstitial solution and the clay component of sediments. Changes in the ionic composition of the interstitial water were first recognized during deep-sea sediment investigations [34].

Cation exchange is one of the first halmyrolytic reactions to occur when clays enter sea water: sodium and magnesium ions are taken up, and calcium and potassium ions are released. Carroll and Starky [35], who studied the titration behavior of hydrogen-form clays neutralized with sea water, found simple titration behavior for kaolinite. Interlayer clay minerals exhibited more complex behavior associated with neutralization of the hydrogen ions and replacement in the exchange sites by metallic cations. As Russell [36] has shown, river clays react with sea water only to the extent of their ion-exchange capacity. The behavior of calcium, potassium, and sodium ions depends on their concentration in sea water and on the properties of the clay mineral. The ion-exchange reaction accompanying a two week long exposure of a montmorillonite river clay to sea water can be described using a mass action expression [9]. Longer exposures to sea water result in a loss of exchange capacity but do not produce changes in x-ray diffraction patterns. This loss may arise from a minor structural modification of the clay which leads to significant alterations in the exchange properties. Decrease in capacity probably reflects loss of exchange sites and appears to be a common phenomenon for sea water-clay interactions.

Ion-exchange processes in sediments have been monitored by several other workers [37-40], and Lerman and Lietzke [41] have used ion-exchange distribution coefficients from these studies to calculate uptake and migration of ^{90}Sr and ^{137}Cs in lake sediments.

III. RELATIONSHIP BETWEEN CLAY MINERAL STRUCTURE AND ION-EXCHANGE PROPERTIES

Ion-exchange properties of clay minerals are directly related to their structure. Clays are fine-grained minerals built up from sheets or chains of tetrahedrally coordinated (Si, Al, Fe^{3+}) and octahedrally coordinated (Al, Fe^{3+}, Fe^{2+}, Mg) cations. For layered silicates the basic structural units are: (a) silica sheets, composed of SiO_4^{4-} tetrahedra, and (b) gibbsite or brucite sheets, containing aluminum or magnesium ions octahedrally coordinated to hydroxide ions [3,9,15,42].

The aluminosilicate matrix of clay minerals usually consists of parallel layers, each containing at least one sheet of silicate tetrahedra and one of aluminate octahedra. Exchangeable cations are located on the surfaces of these layers. The arrangement of these layers, their dispersion, the frequency and site of isomorphous substitution (replacement of a cation by one of lower charge), and their surface properties govern the ion-exchange behavior of the clays [3,9,43].

Clay minerals composed of aluminosilicate layers are classified according to the combinations and composition of the tetrahedral and octahedral sheets in their component layers. Pure clay minerals are composed of one repeating type of layer, while mixed-layer clays [44] are made of interstratified layers of two or more pure clay types. Mixed-layer clays are generally characterized in terms of type and proportion of component pure clays [45]. The most abundant mixed-layer clay is illite-montmorillonite.

Detailed descriptions of clay mineral structure can be found in standard reference texts [15,20,42,46-48]. Berner [9] discusses clay mineral structure with emphasis on its role in sediment geochemistry.

A. Structure - Ion-Exchange Relationship for Kaolinite

The simplest clay mineral is kaolinite (Fig. 1A, Table 1), which consists of stacked layers, each containing a silicate sheet and an aluminate sheet [49]. Although no isomorphous substitution occurs in either sheet, at the edges and corners of the layers are free hydroxyl groups which can particpate in both cation and anion exchange. The forces between layers are relatively weak, allowing easy dispersion of the mineral in water. Ion-exchange rates for kaolinite are high because of this facile dispersion and the location of exchange sites on the clay particle surface, but ion-exchange capacity can vary because of flocculation, which may occur even at moderate salt concentrations.

B. Structure - Ion-Exchange Relationship for More Complex Clays

More complex ion-exchanging clay minerals have layers consisting of one aluminate sheet sandwiched between two silicate sheets. Isomorphous substitution is common and may occur in either the octahedral or tetrahedral sheets, depending on the ionic size and

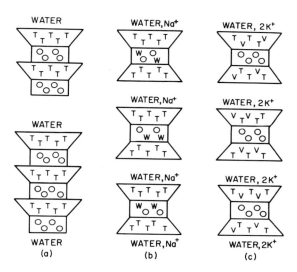

FIG. 1. Schematic illustration of several idealized clays: (a) kaolinite; (b) montmorillonite; (c) illite. T indicates tetrahedrally coordinated sites: V, charge vacancies at tetrahedrally coordinated sites; O, octahedrally coordinated sites; and W, charge vacancies at octahedrally coordinated sites.

coordination number of the substituted ion relative to aluminum and silicon. As a result of this substitution, the aluminosilicate matrix acquires a negative charge, and cations are incorporated in the spaces between the layers to preserve electroneutrality. The cation-exchange capacity of these minerals is due to these mobile cations, which can exchange with cations present in solution when the clay is suspended in an aqueous solution. Additional exchange sites may exist at edges and corners of the clay particles; however, their contribution to the total capacity of the interlayer clays is usually small [50].

The extent of clay dispersion in solution influences the kinetics of ion exchange in natural waters. The dispersion is regulated by the site of isomorphous substitution. Substitution in the octahedral sheets in clays such as montmorillonite increases the strength of bonding between aluminosilicate layers through electrostatic interaction between the layers and the interlayer

TABLE 1

Clay Mineral Ion-Exchange Materials

Mineral	Idealized unit cell content or chemical formula	Ref.	Capacity (meq/100 g)	Ref. [15]
Fibrous				
Attapulgite			10-35	(p.123)
Sepiolite	$Mg_9Si_{12}O_{30}(OH)_6(OH_2)_4 \cdot 6H_2O$	[15, p.127]	20-45	(p.130)
Palygorskite	$Mg_5Si_8O_{20}(OH)_2(OH_2)_4 \cdot 2H_2O$	[15, p.123]	20-36	(p.120)
Laminar				
Glauconite	$Na_2Mg_4Si_7AlO_{20}(OH)_4 \cdot nH_2O$	[42, p.207]	5-40	(p.38)
Halloysite	$Al_4Si_4(OH)_8O_{10} \cdot 8H_2O$	[42, p.257]	60	(p. 152)
Illite	$KAl_4(Si_7AlO_{20})(OH)_4$	[42, p.260]	10-40	(p.19)
Kaolinite	$Al_4(Si_4O_{10})(OH)_8$	[42, p.255]	3.6-18	(p.143)
Smectite				
Montmorillonite	$Na_{0.66}Al_4Si_{7.34}Al_{0.66}O_{20}(OH)_4 \cdot nH_2O$	[42, p.266]	70-130	(p.69)
Saponite	$Na_{0.66}Mg_6Si_{7.34}Al_{0.66}O_{20}(OH)_4 \cdot nH_2O$	[42, p.266]	76	(p.143)
Vermiculite	$Ca_{0.7}Al_6((Al,Si)_8O_{20})(OH)_4 \cdot 8H_2O$	[42, p.270]	80-200	(p.101)
Amorphous				
Allophane			70-100	(p.157)

cations. Because of this interaction, the clays do not readily disperse, but they do swell on immersion in an electrolyte solution. Cation size within the interlayer space regulates the extent of interlayer swelling, which is reversible except under some conditions of high water uptake. The interlayer expansion of the octahedrally substituted clays in aqueous solution facilitates the diffusion of ions into interlayer spaces, thus making rates of ion exchange fairly rapid.

Clay minerals such as illite, with substitution in the tetrahedral sheets, characteristically expand very little when immersed in an electrolyte solution as a result of electrostatic forces between the interlayer cations and the proximate negatively charged tetrahedral sheets. This lack of expansion reduces the diffusion rate of the ions in the interlayer region, leading to very slow rates of ion exchange. Because of the rigid steric limitations imposed by the nonexpandable lattice, large unhydrated cations may be excluded from the mineral.

After initial uptake of ions such as potassium, ammonium [51-53], rubidium [54,55], and cesium [56,57], clay minerals with tetrahedral-sheet isomorphous substitution lose some of their ion-exchange capacity. It has been suggested that this effect is due to fixation, involving slow penetration of the cations from the interlayer space to the tetrahedral layer, where they become unavailable for ion exchange. If peripheral layers collapse when an ion is adsorbed (decreasing the interlayer spacing), larger cations will be trapped in the clay matrix.

IV. ION-EXCHANGE EQUILIBRIA AND SELECTIVITY EXPRESSIONS FOR COMPONENTS OF NATURAL EXCHANGE SYSTEMS

To apply ion-exchange principles to natural water systems it is first necessary to know the affinity of each of the ions in natural waters for the major mineral exchangers. Although affinity series have been developed for several exchange materials, extensive data for affinities over the range of natural conditions are for the most part unavailable.

In order to compare selectivity measurements obtained from experiments conducted under different conditions of temperature, ionic strength and composition, we must devise a unified approach to the calculation of ion-exchange selectivities, one which takes into account all possible variations within a heterogeneous system undergoing an ion-exchange reaction.

A thermodynamic approach to the expression of selectivity results has recently been applied to ion exchangers where exchange sites are not equally available to each ion. It has been used to calculate the selectivity over a range of solution concentrations.

A. Affinity Series for Ions on Natural Ion-Exchange Materials

Eisenman, in an analysis of ion-exchange selectivity in glasses, concluded that the primary physical variable controlling ion-exchange selectivity is the anionic field strength of the exchange site. His approach leads to 11 selectivity sequences for the cations lithium, sodium, potassium, rubidium, and cesium instead of the 120 sequences expected for random behavior [58]. Sherry [59] has used the approach of Eisenman [58] to analyze ion exchange in zeolites.

Barrer and Klinowski [60] have recently discussed affinity series for several different ion-exchange materials. Ion-exchange affinity of a number of zeolites, where ion-sieving properties are known to be absent, was found to be according to hydrated ionic radius. For monovalent ions the ion-exchange affinity sequence is $Li^+ < Na^+ < K^+ < Rb^+ < Cs^+$, while for divalent ions it would be $Mg^{2+} < Ca^{2+} < Sr^{2+} < Ba^{2+}$. This regularity appeared only when exchanges which did not go to completion were normalized by considering the equilibria only on terms of the fraction of ion available for the exchange reaction. The selectivity series obtained shows that the zeolite favors the least hydrated ion while the solution phase favors the most highly hydrated ion.

Ion-exchange affinity series in some clay minerals are complicated by ion fixation. Affinity of clays for hydrated hydrogen ions, although of major importance in view of the acidic rainfall occurring in many areas of the world, is difficult to determine, since treatment of clays with hydrogen ion results in a disruption of the aluminosilicate backbone of the minerals [61].

B. Experimental Measurements Used to Characterize Ion-Exchange Equilibrium and Selectivity for Natural Materials

Ion-exchange capacity is expressed as milliequivalents per 100 g of dry material in a specified ion form [2]. Cation-exchange capacity measurements for clay minerals are commonly obtained by exchange of a specific salt form with 1 N ammonium acetate solution at pH 7 [10]. An alternate method involves a ^{137}Cs-exchange procedure. A detailed discussion of ion-exchange capacity has been presented by Helfferich [2]. For weakly acidic or basic ion-exchange functional groups the capacity is a function of pH. As will be shown subsequently (Section VC; Table 9), ion-exchange capacity also varies according to the ionic composition of the clay mineral. The capacity of clays for cations decreases in the order montmorillonite > illite > kaolinite.

The ion-exchange distribution coefficient is a measure of the extent to which an ion has been removed from solution during exchange. It is defined as the ratio of adsorption to concentration, i.e., the ratio of the number of milliequivalents of an ion adsorbed per gram of exchange material to the number of milliequivalents of that ion per milliliter of solution at equilibrium [2,3]. Jenne and Wahlberg [8] and Duursma and Gross [62] used a similar definition in their description of the distribution of trace metals between natural waters and sediments.

The ion-exchange selectivity coefficient is a measure of the preference of the ion-exchange material for a particular ion. It corresponds to the mass action law coefficient for the exchange reaction and is calculated as the ratio of the concentrations of the two ions in the solid phase divided by that of their concentrations in the solution phase at equilibrium with ionic valences as exponents.

In considering ion-exchange processes in clay minerals, it is convenient to express the results of selectivity measurements as isotherms. For a reaction in which the exchanging ions are $A^{Z_A^+}$ and $B^{Z_B^+}$, having charges Z_A^+ and Z_B^+ respectively, the ion-exchange isotherms are typically expressed as plots of the equivalent cation fraction, X_A, of the ion $A^{Z_A^+}$ in solution against the fraction $\overline{X}_A$ of the same ion in the ion-exchanging clay mineral. Often the parameters X_A and $\overline{X}_A$ are called "relative concentration" and "relative uptake." The equivalent fraction of ion A in solution is calculated from the ratio of the number of equivalents of A per liter to the solution total normality (TN). In the exchanger phase the equivalent fraction of ion A is determined from the ratio of the number of milliequivalents of ion A per 100 g of dry exchanger to the capacity of the exchanger. A typical isotherm for ion-exchange on montmorillonite is shown in Fig. 2.

C. Treatment of Ion-Exchange Equilibria and Selectivity

Approaches characterizing the equilibrium between an ion-exchange material and an aqueous solution have been based on (a) empirical adsorption isotherms, (b) the Donnan equilibrium concept, (c) the diffuse double-layer theory, or (d) the law of mass action.

Helfferich has summarized the application of adsorption isotherms [63] and the Donnan equilibrium concept [64] to ion-exchange processes. Several recent publications use expressions developed from the diffuse double-layer theory to delineate ion-exchange interactions in soils. Lagerwerff and Bolt [65] state that calculations using Gapon's equation based on the diffuse

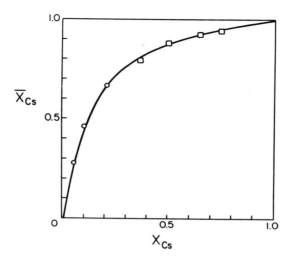

FIG. 2. Isotherm for cesium-ammonium ion exchange on a lower greensand soil: O, 0.010 N; ☐, 0.02 N. Data replotted from Ref. [3].

double-layer theory demonstrate that the exchange factor is constant for conditions such as those encountered in natural soils. Examination of their experimental data, however, shows that the exchange constant varies with changes in solution and exchanger composition. The Gapon equation has also been applied in two recent publications [66,67].

Argesinger et al.[68], Ekedahl et al. [69], and Högfeldt et al. [70] applied the law of mass action to ion-exchange processes using simplified thermodynamic arguments. In 1953 Gaines and Thomas [6] proposed a rigorous thermodynamic treatment: they calculated a thermodynamic equilibrium constant for an ion-exchange reaction employing a model defined in terms of experimental variables. The solvent was treated as an independent variable, allowing all components in the ion-exchange reaction to be taken explicitly into account. Thomas and collaborators [6,7,56,71-73] used well-characterized reference clay minerals supplied by the American Petroleum Institute in mixture with non-exchanging asbestos in packed columns. They examined clay transformations to determine the saturation capacity of the mineral, monitored exchange between the cesium and other ionic forms of the clay to demonstrate exchange reversibility, and studied the ion-exchange equilibrium of two competing cations to determine their distribution coefficient and selectivity coefficient [3]. Cruickshank and Meares [74] used the triangle rule to test the validity of the approach developed by Thomas and collaborators for the ion triplets sodium-cesium-potassium [72] and sodium-cesium-barium [73].

The thermodynamic treatment of Gaines and Thomas appears to have the greatest utility in analysis of ion-exchange components of natural waters. Ion-exchange selectivity measurements obtained over the entire range of solid phase composition are employed in arriving at a thermodynamic ion-exchange selectivity coefficient. This work has been extended recently to include incomplete exchange and selectivity variation with TN.

D. Development of Expressions for Presenting Experimental Ion-Exchange Selectivity Data

Ion exchange between ion A^{Z_A+} in solution and B^{Z_B+} on a clay ion-exchange material has the form

$$Z_A \overline{B}^{Z_B+} + Z_B A^{Z_A+} \rightleftharpoons Z_A B^{Z_B+} + Z_B \overline{A}^{Z_A+} \qquad (1)$$

where the barred terms refer to the ion-exchanger phase. The thermodynamic equilibrium constant or thermodynamic ion-exchange selectivity constant for this reaction is

$$K^A_{a_B} = \frac{\{\overline{X}_A\}^{Z_B} \{a_B\}^{Z_A}}{\{\overline{X}_B\}^{Z_A} \{a_A\}^{Z_B}} = \frac{\overline{X}_A^{Z_B} (m_B)^{Z_A} f_A^{Z_B} \gamma_B^{Z_A}}{\overline{X}_B^{Z_A} (m_A)^{Z_B} f_B^{Z_A} \gamma_A^{Z_B}} \qquad (2)$$

where $\{\overline{X}\}$ represents rational activity in the clay ion-exchanger phase and $\{a\}$ is molal activity in the solution phase; m_A and m_B

are molalities of ions A and B in solution, with solution-phase activity coefficients γ_A and γ_B; and f_A and f_B are rational activity coefficients for the ions associated with the ion-exchanging clay mineral.

Rational activities in the clay are defined so that the standard and reference states for each component are the pure ionic forms. Solution-phase molal activities are defined in the standard way. Helfferich discusses the use of the molal and the rational scales for calculating an ion-exchange equilibrium constant [2,75].

The ion-exchange selectivity coefficient is

$$K_{c_B}^A = \frac{\overline{X}_A^{Z_B} (m_B)^{Z_A}}{\overline{X}_B^{Z_A} (m_A)^{Z_B}} \tag{3}$$

where $K_{c_B}^A$ indicates the experimentally observed extent to which the ion A^{Z_A+} can displace B^{Z_B+}. In solution the single-ion activity coefficient ratio $\gamma_B^{Z_A}/\gamma_A^{Z_B}$ may be replaced by the mean ionic activity coefficient ratio

$$\gamma_{A \pm AX}^{Z_A}/\gamma_{B \pm BX}^{Z_B}$$

when a common anion is employed. Data for mean ion activity coefficients are available for simple salt solutions; in mixed salt solutions standard methods [76] may be applied to correct for ion-ion interaction. Combining Eq. 2 and 3

$$K_{a_B}^A = K_{c_B}^A \frac{f_A^{Z_B} \gamma_B^{Z_A}}{f_B^{Z_A} \gamma_A^{Z_B}} \tag{4}$$

Solution-phase activity coefficients can be incorporated directly into the selectivity expression, giving the modified selectivity coefficient (Kielland) quotient

$$K_B^A = \frac{\overline{X}_A^{Z_B} (m_B)^{Z_A} \Gamma}{\overline{X}_B^{Z_A} (m_A)^{Z_B}} \tag{5}$$

where Γ is equal to $\gamma_B^{Z_A}/\gamma_A^{Z_B}$. Hence

$$K_{a_B}^A = K_B^A \frac{f_A^{Z_B}}{f_B^{Z_A}} \tag{6}$$

With experimental data of the type shown in Fig. 2 and activity coefficients from the literature it is possible to calculate the ion-exchange selectivity coefficient, but additional operations are needed to obtain the thermodynamic ion-exchange selectivity constant.

Other forms for the ion-exchange selectivity coefficient based on the law of mass action have been described, as for example, Nikol'skiy's equation [77] for the exchange reaction

$$KX + \tfrac{1}{2}Mg^{2+} \rightleftharpoons K^+ + \tfrac{1}{2}MgX \tag{7}$$

which gives the selectivity coefficient

$$K_{N_K}^{Mg} = \frac{\overline{X}_{Mg}^{\frac{1}{2}} \cdot X_K}{\overline{X}_K \cdot X_{Mg}^{\frac{1}{2}}} \tag{8}$$

where $\overline{X}_{Mg}$ and $\overline{X}_K$ are the equivalent fractions of magnesium and potassium in the ion-exchanger phase, and X_{Mg} and X_K are the equivalent fractions of the same ions in solution. As will be shown shortly, this formulation of the selectivity coefficient for an unsymmetrical exchange is a function of TN.

The method of Gaines and Thomas [6] for calculation of the thermodynamic ion-exchange selectivity coefficient assumes that a clay mineral has a fixed number of ion-exchange sites permeable to water but not to anions. In general, when exchange involves cations A^{Z_A} and B^{Z_B}, the equilibrium constant is given by the expression [78]

$$\ln K^A_{a_B} = (Z_B - Z_A) + \ln\left[f_A^{Z_B}(a)/f_B^{Z_A}(b)\right] + \int_0^1 \ln K^A_B \, d\overline{X}_A - Z_A Z_B \int_a^b n_s \, d \ln a_s \qquad (9)$$

where n_s is the water content of one equivalent of clay, a_s is water activity, and a and b are the water activities of solutions in equilibrium with pure A and B forms of the exchanger. The first term can be obtained directly. The third term is computed by integration of ($\ln K^A_B \, d\overline{X}_A$) where $\ln K^A_B$ is expressed as a function of $\overline{X}_A$.

Polynomial expressions are used to fit $\ln K^A_B$ to $\overline{X}_A$. Plots of $\ln K^A_B$ versus $\overline{X}_A$ are extrapolated to definite intercepts. This procedure, which assumes that an analog of Henry's law is obeyed in the trace region, has been shown to be appropriate in the case of low concentrations of the more strongly adsorbed ions [73]. For low concentrations of weakly-held ions competition for the exchange sites by hydrogen ions in solution may become significant. In any event, uncertainties in the calculated thermodynamic ion-exchange selectivity coefficient and the corresponding free energies arising from these extrapolations are small [73].

The second term in Eq. 9 represents the change in the pure clay form of the exchanger going from an infinitely dilute solution to the equilibrium solution; the last term is the contribution related to change in water activity in the clay mineral going from a pure B form clay to a pure A form. Amphlett [3] has assumed that the second and last terms are negligible in comparison with the third term (within experimental error), so that

$$\ln K^A_{a_B} = (Z_B - Z_A) + \int_0^1 \ln K^A_B \, d\overline{X}_A \qquad (10)$$

and has presented simplified expressions for the thermodynamic selectivity constant as a function of exchanging-ion charge. A detailed examination of this question by Barrer and Klinowski [79] confirms that the terms can be omitted without significant error.

By employing such simplified equations Thomas and collaborators obtained thermodynamic coefficients for ion exchange on montmorillonite [7,56,71,73] and attapulgite [72]. Natural materials which have been similarly characterized include vermiculite [80,81], greensand (a soil containing montmorillonite and illite) [82,83], and other soils [84]. The results of these investigations will be discussed in Section V.

E. Thermodynamic Treatment of Ion-Exchange Selectivity as a Function of Solution Normality

Ion-exchange selectivity isotherms calculated from the expressions of Gaines and Thomas [6,7] cover a limited concentration range. To extend these results to natural water systems it is necessary to evaluate quantitatively the changes in shape and position of the ion-exchange isotherm with variation in TN of the external electrolyte solution. Such changes would occur, for example, when a suspended clay mineral is carried by a river of low TN to an ocean of high TN. Such analyses have been developed by Barrer and Klinowski [79]. For an exchange reaction such as that shown in Eq. 1, an expression for the ion-exchange selectivity can be written as shown in Eq. 2. If molality m is converted to an equivalent fraction in the solution phase X_A and X_B, then

$$X_A = \frac{Z_A m_A}{Z_A m_A + Z_B m_B} \quad \text{and} \quad X_B = \frac{Z_B m_B}{Z_A m_A + Z_B m_B} \tag{11}$$

and the selectivity coefficient expression becomes

ION-EXCHANGE MATERIALS IN NATURAL WATER SYSTEMS

$$K_{a_B}^A = \frac{\overline{X}_A^{Z_B} X_B^{Z_A}}{\overline{X}_B^{Z_A} X_A^{Z_B}} \cdot \frac{1}{Q} \cdot \frac{f_A^{Z_B} \gamma_B^{Z_A}}{f_B^{Z_A} \gamma_A^{Z_B}} \qquad (12)$$

where

$$Q = Z_B^{Z_A}/(Z_A^{Z_B})(TN)^{(Z_A - Z_B)} \qquad (13)$$

and

$$TN = Z_A m_A + Z_B m_B \qquad (14)$$

Values of Q for various combinations of Z_A and Z_B are given in Table 2.

Evaluation of alterations in the position and shape of a reference isotherm with variations in the external solution TN requires knowledge of changes in Q and of variation in activity coefficients of ions both within the exchanger and in the external aqueous solution.

TABLE 2

Q for Different Z_A and Z_B (Ref. [79])

Z_A	Z_B	Q
1	1	1
2	1	1/2 (TN)
2	2	1
3	1	$1/3 (TN)^2$
3	2	8/9 (TN)
3	3	1

Variation of ion-exchange phase activity coefficients with changing TN can be evaluated employing the treatment of Gaines and Thomas [6,7] to calculate the activity coefficients of ions A and B in the exchanger. Barrer and Klinowski [79] demonstrated that a given value of $\overline{X}_A$ can correspond to different values of the solid-phase activity coefficients depending on the TN of the aqueous solution. The solid-phase activity coefficient ratio at composition p has been given [79] as

$$\ln\,(f_A^{Z_B}/f_B^{Z_A}) = -(Z_A - Z_B) + \int_0^1 \ln\,K_B^A\,d\overline{X}_A - \ln(K_B^A)_p \qquad (15)$$

omitting the terms neglected in Eq. 10.

For a given $\overline{X}_A$ the only physical factors affecting K_B^A are water activities and salt imbibment; these variables do not measurably influence K_B^A for clay ion-exchange materials. Thus, for a given $\overline{X}_A$ the solid-phase activity coefficient ratio should be nearly independent of external electrolyte concentrations, and the following relationship holds:

$$\frac{m_B^{Z_A}}{m_A^{Z_B}}\,\Gamma = \frac{X_B^{Z_A}}{X_A^{Z_B}}\,\Gamma = \frac{K_B^A\,\overline{X}_B^{Z_A}}{\overline{X}_A^{Z_B}} = \frac{K_{a_B}^A\,\overline{X}_B^{Z_A}\,f_B^{Z_A}}{\overline{X}_A^{Z_B}\,f_A^{Z_B}} = \text{constant} \qquad (16)$$

All these ratios are independent of the TN of the external solution for a fixed value of the equivalent fraction of ion A in the ion exchanger.

A convenient separation coefficient, α, is defined [79] as

$$\alpha = \frac{\overline{X}_A\,m_B}{\overline{X}_B\,m_A} = K_{c_B}^A{}^{1/Z_A}\left(\frac{X_A}{m_A}\right)^{(Z_A - Z_B)/Z_A} \qquad (17)$$

where $K_{C_B}^A$ is the selectivity factor. For this equation when $Z_A = Z_B$, $\alpha = K_{C_B}^A{}^{1/Z_A}$ and thus

$$\alpha = \frac{\overline{X}_A (1 - X_A)}{(1 - \overline{X}_A) X_A} \tag{18}$$

When $Z_A \neq Z_B$

$$\alpha = \frac{Z_A \overline{X}_A X_B}{Z_B \overline{X}_B X_A} \tag{19}$$

When $Z_A = Z_B$, Q in Eq. 16 is always unity, and the quotient $X_B^{Z_A}/X_A^{Z_B}$ for a fixed $\overline{X}_A$ can change with TN only as Γ varies. For dilute solutions Γ varies only slightly from unity; therefore, for a fixed value of $\overline{X}_A$, X_A shows little change, and all isotherms must be closely grouped.

When $Z_A \neq Z_B$ for a chosen value of X_A and there are two different normalities, TN_1 and TN_2, then from Eq. 16

$$\frac{X_{A1}^{Z_B} Q_1}{(1 - X_{A1})^{Z_A} \Gamma_1} = \frac{X_{A2}^{Z_B} Q_2}{(1 - X_{A2})^{Z_A} \Gamma_2} = \text{constant} \tag{20}$$

where X_{A1}, Q, and Γ_1 refer to TN_1 and where X_{A2}, Q_2, and Γ_2 refer to TN_2. Changes in TN and thus in Q require a corresponding change in $X_A/(1-X_A)$ through a change in X_A, since the alteration in Γ is usually minor. Thus ion-exchange isotherms (when $Z_A \neq Z_B$) are strongly dependent on the TN of the aqueous solution. As a consequence of the law of mass action for isotherms plotted as $\overline{X}_A$ against X_A, the selectivity of the exchanger for the ion of higher valence as expressed by α, the separation factor (Eq. 19), must increase with dilution. As long as the terms omitted from the integration in Eq. 10 are negligible, the changes in the ion-exchange isotherms with TN can be calculated a priori and do not require any knowledge of the properties of the exchanger other than one measured isotherm for one TN. Details of this calculation are given by Barrer and Klinowski [79].

As an alternative to plots of $\bar{X}_A$ against X_A, which at different TN and for $Z_A \neq Z_B$ produce a three-dimensional surface, it is useful to represent the isotherms at different TN by a single curve. This can be accomplished by plotting $(m_B)^{Z_A} \Gamma / (m_A)^{Z_B}$ or $X_B^{Z_A} \Gamma / X_A^{Z_B} Q$ as a function of $\bar{X}_A$ in the range of validity of the expressions presented in Eq. 16.

An important observation based on these results is that for exchange reactions where $Z_A \neq Z_B$, isotherms which have no inflection point at one dilution can become sigmoid at others. Sigmoid isotherms have frequently been cited as evidence for more than one type of ion-exchange site in an exchanger but only when $Z_A = Z_B$ is it possible to conclude from a sigmoid isotherm the nature or population of several distinct types of sites.

This has important implications for natural water-clay interactions when meteoric water of low TN comes in contact with clay minerals. For exchanges where $Z_A \neq Z_B$, the exchanger exhibits high selectivity for the ion of higher valence in solutions of low TN. Uptake of multivalent ions can arise universally from high dilution of the electrolyte solution independent of the ion-exchanger phase. In the limiting situation of extreme dilution of the aqueous phase, the binary system shown in Eq. 1 may become a ternary system involving A^{Z_A}, B^{Z_B}, and H_3O^+ supplied by the water.

F. Thermodynamic Treatment of Incomplete Exchange

It is essential to examine ion exchange in materials where not all ions are exchangeable. Incomplete exchange is fairly frequent among natural ion-exchange materials, where it may reflect steric interference with the exchange of some ions in a specific region of the mineral.

The ion-exchange treatment of Gaines and Thomas [6,7] defines the standard states for ions A and B in the exchanger as the respective homoionic forms. This definition is appropriate when the exchange reaction proceeds to completion. However, for some exchange reactions it is not possible to prepare a homoionic exchanger for certain cations. For reactions where only a fraction of the ion-exchange sites are available to one of the exchanging ions, the standard state would be one with all available sites occupied by that ion [85-87]. The treatment of Gaines and Thomas has been modified rigorously to accommodate limited exchange-site availability during an ion exchange reaction [8]. A normalized Kielland selectivity coefficient K_B^A is expressed in terms of the equivalent cation fractions $\overline{X}_A$ and $\overline{X}_B$ of the exchangeable cations only. The thermodynamic ion-exchange selectivity constant is then calculated using the expression in Eq. 10 with the integral of the normalized Kielland selectivity coefficient.

V. EXPERIMENTAL MEASUREMENTS OF ION-EXCHANGE SELECTIVITY FOR COMPONENTS OF NATURAL WATER SYSTEMS

Characterization of ion-exchange processes in natural waters has been extended by examining the exchange properties of natural sediments in the laboratory. The findings parallel ion-exchange behavior in the environment. In addition, they serve as a practical demonstration of the types of interactions ions such as radionuclides would exhibit when released into the environment.

A. Selectivity Measurements for Pure Reference Clays

Thomas and his collaborators made extensive investigations of 1:1, 1:2, and 1:3 ion-exchange systems with montmorillonite [7,56,71] and attapulgite [72] and obtained the thermodynamic equilibrium constants and free energies of the exchange reactions (Table 3: see Section IVC for an account of techniques used). Measurement

TABLE 3

Standard Free Energy of Exchange for Two Clay Minerals

Clay mineral	Exchange reaction*	$\Delta G°$ (kcal/mole)	Ref.
Attapulgite	$Li^+ \to Cs^+$	2.14	[3, p.28]
	$Na^+ \to Cs^+$	1.93	
	$K^+ \to Cs^+$	1.22	
Montmorillonite	$K^+ \to Cs^+$	1.5	
	$Sr^{2+} \to Cs^+$	3.7	
	$Y^{3+} \to Cs^+$	4.7	
	$Na^+ \to Cs^+$	2.15	[73]
	$Na^+ \to Ba^{2+}$	0.5	
	$Ba^{2+} \to Cs^+$	3.68	

* $Li^+ + CsX \rightleftharpoons Cs^+ + LiX$

$$K^{Li}_{a_{Cs}} = \frac{\overline{X}_{Li}\, m_{Cs}\, f_{Li}\, \gamma_{Cs}}{\overline{X}_{Cs}\, m_{Li}\, f_{Cs}\, \gamma_{Li}}$$

$\Delta G° = -RT \ln K$

of ion-exchange selectivities as a function of temperature allowed calculation of the enthalpies and entropies of exchange. Ion-exchange isotherms [7,56,71] are presented in Fig. 3 to illustrate graphically the exchange behavior of cesium with monovalent, divalent, and trivalent ions. Amphlett[3] discusses Thomas's results in detail and points out that the experimental isotherms become more complicated as the charge on the exchanging ion increases.

Studies of ion exchange using kaolinite [89] show that the exchanger phase prefers cesium to sodium ion. Cation-exchange capacities varied with the concentration and type of clay used.

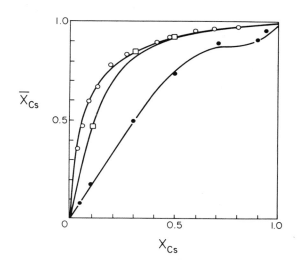

FIG 3. Isotherms for cesium ion exchange on montmorillonite with monovalent, divalent and trivalent ions. O, cesium-potassium, 0.01-0.04 N at room temperature; □, cesium-strontium, 0.05 N at 25° C; ●, cesium-yttrium, 0.04 N at 30° C. Data replotted from Ref. [3].

Slow attainment of ion-exchange equilibrium indicated that the state of aggregation of the clay particles was not constant during these experiments. Flocculation appeared to reduce the availability of cation-exchange sites for the exchange reaction. By analogy, it would be difficult to describe the ion-exchange properties of kaolinite in natural waters (where the degree of flocculation is unknown). The exchange mechanism of kaolinite has recently been reexamined [90], and it has been concluded that the clay does not exhibit a constant ion-exchange capacity because of the presence of an aluminosilicate gel coating on the kaolinite surface.

Other pure clays have been studied in the laboratory by Frysinger and Thomas [91], who experienced additional difficulties in characterizing model natural ion-exchange materials when they examined the ion-exchange behavior of vermiculite-biotite mixtures. They found that the exchange reaction was complicated by a slow conversion of biotite to vermiculite. Conversion occurred most readily at high ionic strengths and temperatures. Such ion-exchange material in natural systems would probably show a time dependent selectivity behavior related to the ionic composition of the surrounding electrolyte solution.

Using American Petroleum Institute reference clays kaolinite No. 4, kaolinite No. 7, illite No. 35, and montmorillonite No. 21, Wahlberg et al. [92] have determined distribution coefficients for the exchange of strontium with the major cations in natural waters (magnesium, calcium, sodium, and potassium). To reduce the complexity of the system, amorphous aluminum and iron oxides and readily acid-soluble minerals were removed from the clay exchange materials prior to measurement of distribution coefficients. Distribution coefficients were measured using a dialysis tube containing the metal form of the clay and an exchanging solution containing a known strontium concentration.

For the exchange reaction

$$CaX_2 + Sr^{2+} \rightleftharpoons SrX_2 + Ca^{2+} \tag{21}$$

the distribution coefficient was defined by Wahlberg as

$$K_d = (SrX_2)/(Sr^{2+}) \tag{22}$$

where SrX_2 represents the amount of strontium adsorbed by the clay, in meq/g, and Sr^{2+} is the normality of strontium in the equilibrium solution. The mass action expression employed to analyze the data is

$$K'^{Sr}_{Ca} = \frac{(SrX_2)(Ca^{2+})}{(Sr^{2+})(CaX_2)} = K_d \frac{(Ca^{2+})}{(CaX_2)} \tag{23}$$

where (CaX_2) is the amount of calcium adsorbed by the clay, in meq/g, (Ca^{2+}) is the normality of calcium in the equilibrium

TABLE 4

Ion-Exchange Selectivity Coefficients and Cation-Exchange Capacity for Several American Petroleum Institute (API) Reference Clay Minerals (Ref. [92])

Clay (API reference clay number)	$K'^{Sr}_{c_M}$	$K'^{H}_{c_M}$	Cation exchange capacity (meq/g)
Potassium montmorillonite No. 21	0.2	1.0	1.07
Sodium montmorillonite No. 21	1.1	10.0	1.07
Magnesium montmorillonite No. 21	1.1	—	1.07
Calcium montmorillonite No. 21	1.1	—	1.07
Potassium illite No. 35	1.0	14.0	0.28
Sodium illite No. 35	3.0	10.0	0.28
Magnesium illite No. 35	1.5	—	0.145
Calcium illite No. 35	1.1	—	0.145
Potassium kaolinite No. 4	1.5	2.7	0.133
Sodium kaolinite No. 4	4.2	8.3	0.190
Magnesium kaolinite No. 4	1.0	—	0.246
Calcium kaolinite No. 4	0.9	—	0.246
Potassium kaolinite No. 7	1.5	23.0	0.056
Sodium kaolinite No. 7	9.0	30.0	0.056
Magnesium kaolinite No. 7	1.8	—	0.043
Calcium kaolinite No. 7	1.2	—	0.043

solution, and $K'^{Sr}_{c_{Ca}}$ represents the mass-action equilibrium coefficient.

Ionic concentrations were used in the calculation of the mass-action equilibrium coefficient, which for strontium exchange with either calcium or magnesium was nearly unity and did not vary with exchange capacity (Table 4). In contrast, for exchange of sodium and potassium with strontium the mass-action coefficient increased as the exchange capacity decreased.

The results presented by Wahlberg et al. are of limited utility because there was no verification of equilibrium having been reached and no demonstration of reversibility for the exchange reactions studied. The exchange reactions were not studied over a wide enough range of exchanger-phase composition to allow valid comparison with other published data. However, the data of Wahlberg et al. do enable estimation of the uptake of trace metal species in natural waters by the ion-exchange material.

In dealing with ion exchange involving trace metals, it is important to recognize that sodium, potassium, calcium, and magnesium, the major cations involved in ion-exchange reactions in natural water, exist primarily in solution as the hydrated free ion. Recent advances in computer assisted calculations, as well as information concerning the stability of ionic complexes in solution, enable calculation of the extent of ion-pair formation under most natural conditions. In analyzing trace-metal ion exchanges it is essential to consider ionic species distribution.

B. Selectivity Measurements for Purified Natural Materials

Wild and Keay [80] have examined the ion-exchange behavior of sodium, magnesium, calcium, strontium, and barium ions on vermiculite in the temperature range 25-70° C. The cation-exchange capacity of vermiculite in the magnesium form was found to be 133 meq/100 g. Chemical and x-ray diffraction analyses were employed to characterize the exchanger material. X-ray analysis revealed the sample to be vermiculite with a slight interstratification of biotite. A batch-equilibration procedure was used for selectivity measurements. Equilibrium was verified for the magnesium-sodium exchange by approaching it from both directions. Ion-exchange selectivity measurements for sodium ion with the alkaline earth metal ions were made using 0.1 N NaCl at 25° and 70°C. As seen in Fig. 4, divalent ions have a greater affinity than sodium ion for the solid phase, in the sequence $Mg^{2+} > Ca^{2+} > Sr^{2+} > Ba^{2+}$. At higher temperatures the selectivity for multivalent ions over sodium ion is greatly increased.

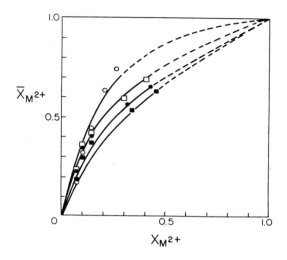

FIG 4. Isotherms for alkaline earth-sodium ion-exchange in 0.1 N solution on vermiculite at 25°C. O, sodium-magnesium; □, sodium-calcium, ●, sodium-strontium; ■, sodium-barium. Data from Ref. [80].

When barium exchange with magnesium, calcium, and strontium was examined in a solution containing 0.020 N $BaCl_2$ at 25° C, the selectivity sequence observed for the multivalent ions was the same (Fig. 5).

For the divalent-monovalent exchange reaction

$$NaX + \tfrac{1}{2}M^{2+} \rightleftharpoons \tfrac{1}{2}MX + Na^+ \tag{24}$$

Wild and Keay [80] used the equilibrium constant expression in the form

$$K^{'M^{2+}}_{a_{Na}} = \frac{N_{M^{2+}}^{\frac{1}{2}} \cdot m_{Na^+} \cdot f_{M^{2+}}^{\frac{1}{2}} \cdot \gamma_{Na^+}}{N_{Na} \cdot m_{M^{2+}}^{\frac{1}{2}} \cdot f_{Na} \cdot \gamma_{M^{2+}}^{\frac{1}{2}}} \tag{25}$$

where $N_{M^{2+}}$ and N_{Na} are the mole fractions of each ion in vermiculite, $m_{M^{2+}}$ and m_{Na} are the molalities of each ion in the solution phase, $f_{M^{2+}}$ and f_{Na} represent the rational activity coefficients in

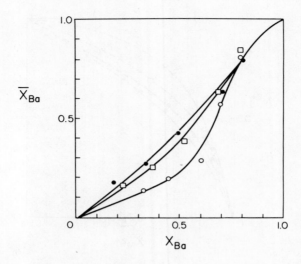

FIG. 5. Isotherms for barium-alkaline earth ion exchange in 0.020 N solution on vermiculite at 25° C. O, barium-magnesium; ☐, barium-calcium, ●, barium-strontium. Data from Ref. [80].

vermiculite (defined as unity in pure M^{2+} and Na^+ vermiculite respectively at unit water activity), and $\gamma_{M^{2+}}$ and γ_{Na} are the molal activity coefficients of the chlorides in the solution phase. Thermodynamic ion-exchange selectivity coefficients were calculated using a simplified form of the expression developed by Gaines and Thomas [6,7]

$$\ln K'^{M^{2+}}_{a_{Na}} = \int_0^1 \ln K^{M^{2+}}_{a_{Na}} \, d\overline{X}_{Na} \qquad (26)$$

where

$$K^{M^{2+}}_{Na} = \frac{N_{M^{2+}}^{\frac{1}{2}} \cdot m_{Na^+} \cdot \gamma_{Na^+}}{N_{Na} \cdot m_{M^{2+}}^{\frac{1}{2}} \cdot \gamma_{M^{2+}}^{\frac{1}{2}}} \qquad (27)$$

and $\overline{X}_{Na}$ is the equivalent fraction of sodium in the vermiculite.

Standard free energies for the exchange reactions were calculated from the equilibrium constants (Table 5). Arrhenius

TABLE 5

Thermodynamic Ion-Exchange Selectivity Constants and Standard Thermodynamic Parameters for Ion-Exchange Reactions on Vermiculite (Ref. [80])

Exchange reaction*	Temp. (°C)	K'_a	$\Delta G°$ (cal/eq)	$\Delta H°$ (cal/eq)	$T\Delta S°$ (cal/eq)
$Mg^{2+} \rightleftharpoons 2Na^+$	25	1.32	−162	+4,800	+4,900
	40	2.12	−467	+4,800	+5,200
	56	2.84	−680	+4,800	+5,400
	70	3.88	−921	+4,800	+5,700
$Ca^{2+} \rightleftharpoons 2Na^+$	25	0.99	+6	+4,700	+4,700
	47	1.88	−401	+4,700	+5,100
	70	2.81	−702	+4,700	+5,400
$Sr^{2+} \rightleftharpoons 2Na^+$	25	1.01	−8	+4,400	+4,400
	47	1.93	−416	+4,400	+4,800
	70	2.70	−674	+4,400	+5,100
$Ba^{2+} \rightleftharpoons 2Na^+$	25	1.01	−6	+5,300	+5,300
	47	1.66	−321	+5,300	+5,700
	70	3.34	−820	+5,300	+6,200
$Ba^{2+} \rightleftharpoons Mg^{2+}$	25	0.73	+185	+600	+400
	70	0.84	+122	+600	+500
$Ba^{2+} \rightleftharpoons Ca^{2+}$	25	0.94	+36	+600	+600
	70	1.08	−55	+600	+700
$Ba^{2+} \rightleftharpoons Sr^{2+}$	25	0.99	+4	+500	+500
	70	1.10	−64	+500	+500

* $\tfrac{1}{2}Mg^{2+} + NaX \rightleftharpoons Na^+ + \tfrac{1}{2}MgX_2$

$$K'^{Mg}_{a_{Na}} = \frac{\overline{X}_{Mg}^{1/2} \, m_{Na} \, f_{Mg}^{1/2} \, \gamma_{Na}}{\overline{X}_{Na} \, m_{Mg}^{1/2} \, f_{Na} \, \gamma_{Mg}^{1/2}}$$

$\Delta G° = -RT \ln K^{Mg}_{a_{Na}}$

$\Delta G° = \Delta H° - T\Delta S°$

plots (log K versus 1/T) were used to determine the standard enthalpy of exchange, which was employed to calculate the entropy of exchange (Table 5). The somewhat greater affinity of magnesium ion compared to the other divalent ions studied was attributed to its interaction with the silicate surface of the vermiculite.

Wilhelm and Wey [81] studied the exchange of sodium and lithium ions on a vermiculite exchanger at 25°, 80°, and 300° C in 0.05 and 0.02 N solutions. The vermiculite was characterized by chemical analysis (Table 6), by x-ray powder diffraction analysis, and by differential thermal analysis. Ion-exchange reversibility was verified by approaching equilibrium from both directions. The capacity of the vermiculite sample was 80 meq/100 g. Results of this work are given in Fig. 6. The thermodynamic ion-exchange selectivity constant calculated from these data (Table 7) is independent of the path to equilibrium within experimental error.

TABLE 6

Chemical Analysis of Vermiculite Sample Used for Ion-Exchange Selectivity Measurements (Ref. [81])

Oxide	SiO_2	Al_2O_3	Fe_2O_3	TiO_2	MgO	Na_2O	K_2O	H_2O
Weight (%)	35.50	12.00	5.35	0.94	27.90	0.20	5.26	13.50

C. Selectivity Measurements for Natural Materials

Deist and Talibudeen [84] employed radioisotopes to monitor ion-exchange equilibria between washed dispersed soils (Table 8) and mixed 0.01 N chloride solutions at 25° C. Cation-exchange capacities (CEC) were determined by using ammonium acetate (Table 8) and by isotopic exchange (Table 9). A decrease in CEC in the calcium-potassium system with increasing potassium equiva-

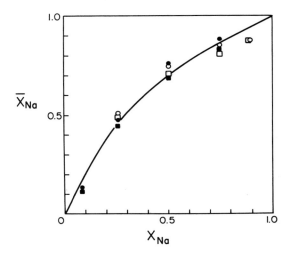

FIG. 6. Isotherms for sodium-lithium exchange on vermiculite at 80°C and at 0.05 N (circles) and 0.02 N (squares). Solid points indicate equilibrium approached from the lithium form of the exchanger; open points indicate equilibrium approached from the sodium form. Data replotted from Ref. [81].

TABLE 7

Thermodynamic Ion-Exchange Selectivity Constant for a Sodium-Lithium Exchange Reaction on Vermiculite at 80° C (Ref. [81])

Solution normality	Exchange reaction*	K_a
0.02	$XLi + Na^+ \leftrightharpoons NaX + Li^+$	2.01
	$XNa + Li^+ \leftrightharpoons LiX + Na^+$	1/2.03
0.05	$XLi + Na^+ \leftrightharpoons NaX + Li^+$	2.23
	$XNA + Li^+ \leftrightharpoons Na^+ + XLi$	1/2.33

* $K_{a_{Li}}^{Na} = \dfrac{\overline{X}_{Na} \, m_{Li} \, f_{Na} \, \gamma_{Li}}{\overline{X}_{Li} \, m_{Na} \, f_{Li} \, \gamma_{Na}}$

TABLE 8

Description of Soils Used by Deist and Talibudeen
for Ion-Exchange Selectivity Measurements (Ref. [84])

Soil	Cation exchange capacity (meq/100 g)	Percent clay (< 2 μ) in soil	Percent fine clay (< 0.2 μ) in total clay
Cegin	15.2	12.6	12.7
Bovey Basin	18.3	17.1	31.0
Tedburn	12.4	24.5	24.5
Dunkeswick	16.6	25.7	41.0
Long Load	30.1	27.6	43.0
Newchurch	22.0	41.4	45.7
Windsor	15.3	22.5	46.6
Denchworth	19.4	42.6	51.4
Sherborne	15.9	29.9	57.9
Harwell	30.7	37.3	63.8

lent fraction started at about 0.3 equivalent fraction of potassium in the exchanger phase. Almost no ion-exchange isotherms obtained in the study showed hysteresis or selectivity reversals. Representative exchange isotherms for the exchange of potassium with sodium, rubidium, and calcium are presented in Fig. 7. Potassium-calcium exchange on Harwell soil was the only exchange reaction which exhibited a selectivity reversal (Fig. 8).

Ion-exchange reversibility was examined by Deist and Talibudeen for all exchange reactions by approaching equilibrium from both sides. In most of the soils tested the uptake of ions was reversible. However, three soils showed hysteresis for potassium-calcium exchange (Fig. 9), involving stronger potassium binding to a potassium-form soil. The results of irreversible potassium binding are seen in the isotopic CEC measurement shown in Table 9.

TABLE 9

Cation Exchange Capacities obtained by Isotopic Exchange for
Several Sodium, Potassium, and Calcium-Saturated
Soils (meq/100 g)(Ref. [84])

Soil series	Cation exchange capacity (meq/100 g)		
	Na	K	Ca
Tedburn	14.1	11.9	16.6
Bovey Basin	13.9	12.2	17.7
Cegin	15.0	12.5	16.0
Windsor	19.4	17.4	22.3
Dunkeswick	19.5	14.5	22.0
Sherborne	-	21.0	26.5
Long Load	-	27.0	33.0
Denchworth	-	19.0	33.0
Harwell	-	34.2	44.2
Newchurch	-	16.0	35.0

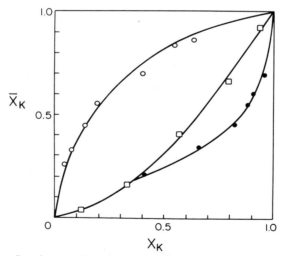

FIG. 7. Isotherms for potassium ion-exchange with sodium, rubidium and calcium ions in 0.01 N solution on a soil. O, potassium-sodium; □, potassium-rubidium; ●, potassium-calcium. Data replotted from Ref. [84].

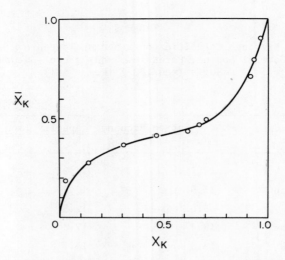

FIG. 8. Isotherm for potassium-calcium exchange on Harwell soil showing a selectivity cross-over at $\overline{X}_K = 0.4$. Data replotted from Ref. [84].

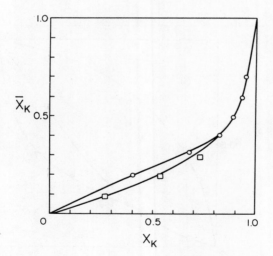

FIG. 9. Isotherm with irreversible hysteresis for potassium-calcium ion-exchange on a soil. O, Ca^{2+} added to the potassium form of the soil; □, K^+ added to the sodium form of the soil. Data replotted from Ref. [84].

These results suggest complications associated with a detailed examination of ion-exchange materials resembling those in natural systems. The variable ion-exchange capacity of several soils undergoing transformation from the calcium to potassium form may be a result of potassium ion fixation in the clay lattice or other subtle structural changes brought about by modification in the composition of the exchanger. The mechanism causing a decrease in the CEC of soil involved in a potassium-calcium exchange was studied by Deist and Talibudeen in a number of clays by examining both the extent of potassium fixation and the change in CEC with increased potassium loading. Two ion-exchange materials did not fix potassium, but their CEC decreased. The vermiculite showed some potassium fixation, but it was too little to account for the change in the CEC. The loss in CEC thus seems to be due to the trapping of larger hydrated ions within the crystal lattice of 2:1 type minerals as lattice-collapsing ions are adsorbed.

Thermodynamic ion-exchange constants for the reactions studied were calculated by using graphic integration in the simplified expression of Gaines and Thomas [6,7] (Eq. 10), and the standard free energies of the exchange reaction were calculated from the expression

$$\Delta G° = -RT \ln K_{a_A}^K \qquad (28)$$

The results of these calculations (Table 10) follow the selectivity series $Rb^+ > K^+ > Na^+$.

The free energy for the calcium-potassium exchange reaction (Table 10) suggests that the reaction proceeds spontaneously to form the potassium clay, whereas the ion-exchange selectivity isotherm for this exchange (Fig. 7) indicates preferential uptake of the calcium ion. As pointed out in Section IV-E, the increased selectivity seen in the ion-exchange isotherm for an unsymmetrical ion-exchange reaction always favors the more highly charged ion as

TABLE 10

Standard Free Energies of Reaction for Potassium Exchange*
with Calcium, Rubidium, and Sodium Ions on Soils (Ref. [84])

Soil series	ΔG° (cal/mole)		
	$2K^+ \rightarrow Ca^{2+}$	$K^+ \rightarrow Rb^+$	$K^+ \rightarrow Na^+$
Tedburn	-1370	+447	-927
Bovey Basin	-1245	+377	-888
Cegin	-1050	+435	-966
Windsor	-1550	+555	-1015
Dunkeswick	-1315	+667	-1083
Sherborne	-2280	-	-
Long Load	-2120	+474	-
Denchworth	-1770	-	-
Harwell	-3420	+565	-
Newchurch	-1900	+515	-

* $CaX_2 + 2K^+ \rightleftharpoons 2KX + Ca^{2+}$

$$K_{a_{Ca}}^{K} = \frac{\bar{x}_K^2}{\bar{x}_{Ca}} \cdot \frac{m_{Ca}}{m_K^2} \cdot \frac{f_K^2}{f_{Ca}} \cdot \frac{\gamma_{Ca}}{\gamma_K^2}$$

$$\Delta G^\circ = -RT \ln K_{a_{Ca}}^{K}$$

the solution TN decreases. As indicated in that section, an isotherm at any TN can be calculated directly from a single ion-exchange isotherm at a fixed TN.

Amphlett and MacDonald [82,83] have studied ion-exchange uptake of cesium and strontium on a lower greensand soil with satisfactory exchange properties. The soil contained little organic matter, and the ion-exchange capacity was stated to be due mainly to its clay mineral content. The clay minerals most abundant in the soil were montmorillonite and illite, with a slight amount

TABLE 11

Thermodynamic Ion-Exchange Selectivity Constants and Standard Free Energies of Exchange for Cesium Ion Exchange* on Lower Greensand at 20°C (Ref. [83])

Exchange reaction	K_a		$\Delta G°$ (cal/mole)
$Sr^{2+} \to Cs^+$	6.27×10^{-4}	(0.02 N)	3900
	6.48×10^{-4}	(0.05 N)	3870
$Na^+ \to Cs^+$	0.057		1850
$NH_3^+ \to Cs^+$	0.128		1190

* $2CsX + Sr^{2+} \rightleftharpoons 2Cs^+ + SrX$

$$K_{a_{Cs}}^{Sr} = \frac{\overline{X}_{Sr}}{\overline{X}_{Cs}^2} \cdot \frac{m_{Cs}}{m_{Sr}} \cdot \frac{f_{Sr}}{f_{Cs}^2} \cdot \frac{\gamma_{Cs}^2}{\gamma_{Sr}}$$

$$\Delta G° = -RT \ln K_{a_{Cs}}^{Sr}$$

of kaolinite. Strontium ion was not fixed by the soil during the exchange, and only slight fixation of cesium was observed. Thermodynamic ion-exchange selectivity coefficients were calculated for the reaction using the approach of Gaines and Thomas [6,7] (Table 11).

In another study, Amphlett examined the exchange of cesium, sodium, and ammonium ion on lower greensand soil (Table 11). The experiments were conducted to accommodate the slow kinetics of exchange. The soil showed less selectivity for cesium than montmorillonite but greater selectivity for this ion than exhibited by attapulgite (Fig. 10). Comparison of the results of exchange on lower greensand with data for montmorillonite, a major component of the soil, shows significant differences, which have been explained on the basis of the presence of illite in the soil. Illite appears to exhibit a lower selectivity for cesium than does montmorillonite.

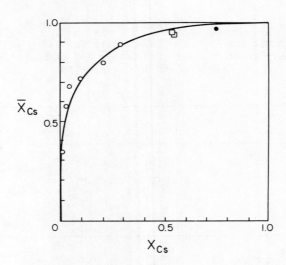

FIG 10. Isotherm for cesium-sodium ion-exchange on a lower greensand soil. O, 0.10 N; □, 0.05 N; ●, 0.02 N. Data replotted from Ref. [3].

The ion-exchange behavior shown by soils according to these publications is similar to that of reference ion-exchange clay minerals. This similarity indicates that an analysis of more complex ion-exchange systems such as sediment might be possible by using the thermodynamic mass-action selectivity expressions applied to soils, even though the problems inherent in analyses of ion-exchange selectivity in soils are compounded in such natural systems. Quantitative characterization of the ion-exchange process may be impeded by the presence of large amounts of organic material, the differing ion-exchange selectivities and rates of reaction of mixed clay minerals, and irreversible ion exchange.

D. Selectivity Measurements for Sediments

Ion-exchange selectivity measurements on reference clay minerals, purified natural minerals, and soils provide an indication of ion exchange behavior even in complex natural systems. Hydrous metal oxides and organic substances are often present in sediments and other natural materials and can act as ion-exchange materials.

Rutkovskiy [77] examined the exchange of calcium and magnesium ion on Pacific Ocean clays in the sodium and potassium form (see Table 12 for characteristics of these clays and Table 13 for their chemical analysis). Thermal analysis of the clay samples showed the presence of hydromica, montmorillonite and amorphous substances.

Ion-exchange selectivities for the clays were calculated by replacing the potassium by magnesium or calcium and then using Nikol'skiy's equation for the potassium-magnesium exchange [77] (Eq. 7 and 8). Table 14 presents ion-exchange selectivity coefficients at several different solution TN that were calculated using Eq. 8. The change in selectivity with solution TN reflects the dependence of the coefficient in Eq. 8 on TN and on solution activity coefficients. The functional relationship is given in Eq. 12 and for the exchange reaction (Eq. 7) yields

$$(K_{N_K}^{Ca})^{\frac{1}{2}} = \frac{K_K^{Ca}}{Q^{\frac{1}{2}}} \Gamma^{\frac{1}{2}} \qquad (29)$$

where K is the selectivity coefficient reported by Rutkovskiy and the other terms are as defined in Eq. 12. Values for Q have been taken from Table 2 and solution-phase activity coefficients from the literature. The results of this calculation (Table 14) illustrate the agreement between selectivities at different TN when Eq. 16 is employed. Selectivities were also determined for exchange reactions at a constant ionic strength. As required by Eq. 16, the results appear to be constant within experimental error. There is good agreement between selectivity measurements for sediment from Station 233 for calcium-potassium exchange when determined in experiments with constant or variable ionic strength.

Jenne and Wahlberg [8] examined the role of ion-exchange processes in stream sediment systems with measurable amounts of

TABLE 12

Characteristics of Two Sediments from the Pacific Ocean (Ref. [77])

Station no.	Depth (m)	Type of sediment	Particle size (%)*				Exchange capacity (meq/100 g)	
			<0.1	0.1-0.5	0.05-0.01	<0.01	Na form	K form
233	4580	Red clay	0.17	0.68	5.92	93.23	48.61	52.92
264	5840	Aleuritic-pelitic mud	2.54	1.97	25.63	69.85	32.04	33.80

* Particle size distribution results are shown as given in Reference [77].

TABLE 13
Chemical Composition (%) of Sediments from the Pacific Ocean (Ref.[77])

Station no.	SiO_2	Al_2O_3	TiO_3	Fe_2O_3	CaO	MgO	K_2O	Na_2O
233	55.60	16.44	0.77	9.92	4.51	5.17	2.10	5.00
264	58.52	19.68	0.50	8.80	3.47	4.27	-	-

illite, using a dialysis-exchange procedure to conduct ion-exchange selectivity measurements. In addition, selectivities were determined by employing a batch-equilibrium process using 2 N ammonium acetate and 2 N ammonium chloride as exchanging solutions.

Selectivity coefficients determined in this study for exchange reactions of the form

$$\tfrac{1}{2}(Ca\overline{X}_2 + Mg\overline{X}_2) + 2Cs^+ \rightleftharpoons 2\overline{CsX} + \tfrac{1}{2}(Ca^{2+} + Mg^{2+}) \quad (30)$$

were approximately 2×10^4 by the batch-equilibrium procedure and 5.6×10^4 for the dialysis-exchange procedure. In these experiments cesium is at trace concentration and the exchanger is in the calcium-magnesium form. Earlier laboratory studies [93] using reference illite sample gave a selectivity value of 4×10^7 for trace levels of cesium in 0.003 N calcium chloride solution. The difference between the selectivities calculated for the reference clay and the sediment has been attributed to possible coatings of organic matter, iron, manganese, and possibly aluminum and silicon on the primary particles, presenting a diffusion barrier. To overcome this problem, equilibration times would need to be considerably increased. In support of this conclusion is the fact that rock fragments require up to twice as much time to reach equilibrium as do the reference clays for ion-exchange reactions [94].

TABLE 14

Ion-Exchange Selectivity Coefficients* for Calcium-Potassium Ion-Exchange on Pacific Ocean Sediments

Station no.	Selectivity coefficient as a function of ionic strength (Ref. [77])		Recalculation using factor given by Barrer (Ref. [79])		
	Ionic strength	$K_{N_K}^{Ca}$	$Q^{1/2}$**	$\Gamma^{1/2}$	K_K^{Ca}
233	0.1	0.27	2.234	1.079	0.130
	0.05	0.35	3.162	1.083	0.120
	0.02	0.65	5.000	1.051	0.137
	0.01	0.70	7.071	1.046	0.103
264	0.05	0.38	-	-	-
	0.02	0.30	-	-	-

* $\frac{1}{2}Ca^{2+} + KX \rightleftharpoons K^+ + \frac{1}{2}CaX$

$$K_{N_K}^{Ca} = \frac{\overline{X}_{Ca}^{\frac{1}{2}}}{\overline{X}_K} \cdot \frac{X_K}{X_{Ca}^{\frac{1}{2}}}$$

** $Q = 1/2$ (TN) The total normality (TN) for the solutions used in the experiments was equal (within 20%) to the ionic strength.

VI. SUMMARY

The importance of ion-exchange materials in natural water systems has yet to be fully recognized, although ion-exchange reactions are often considered in analyses of sediment-water and soil-water interactions. The complex nature of substances encountered in the environment has often restricted more extensive application of ion-exchange principles. Description of most exchangeable species in the solution phase is available using standard thermodynamic procedures, but characterization of the numerous organic and inorganic ion-exchange materials found in the environment presents major difficulties. In addition, interpretation of ion-exchange reactions in natural waters has been hampered by lack of a uniform approach to the calculation of selectivity. Barrer and co-workers [78,82] have recently presented a thermodynamic method that facilitates comparison of selectivity data obtained for a number of systems. Ion-exchange selectivity trends can often be related to fundamental properties of the exchanging ions.

Characterization of natural ion-exchange materials could be used to estimate trace metal ion concentrations in a wide range of natural conditions. Cadmium, copper, lead, and zinc, for example, may be taken up by a natural ion-exchange material near a source of pollution and released in another location by a change in solution composition. Description of these exchange reactions requires accurate ion-exchange selectivity measurements for natural ion-exchange materials at the low trace-metal loadings encountered in the environment.

REFERENCES

1. R. J. Gibbs, Science, 180, 71 (1973).
2. F. Helfferich, Ion Exchange, McGraw-Hill, New York (1962).
3. C. B. Amphlett, Inorganic Ion Exchangers, Elsevier, New York (1964).
4. W.P. Kelly, Cation Exchange in Soils, American Chemical Society Monograph No. 109, Reinhold, New York (1948).

5. B. P. Robinson, Ion Exchange Minerals and Disposal of Radioactive Wastes - a Survey of Literature, U.S. Geol. Surv. Water Suppl. Paper 1616 (1962).
6. G. L. Gaines and H. C. Thomas, J. Chem. Phys., 21, 714 (1953).
7. G. L. Gaines and H. C. Thomas, J. Chem. Phys., 23, 2322 (1955).
8. E. A. Jenne and J. S. Wahlberg, Role of Certain Stream Sediment Components in Radio-Ion Sorption, U.S. Geol. Surv. Prof. Paper, 433-F (1968).
9. R. A. Berner, Principles of Chemical Sedimentology, McGraw-Hill, New York (1971), p. 196.
10. H. E. Rieke, III and G. V. Chilingarian, Developments in Sedimentology 16, Compaction of Argillaceous Sediments, Elsevier, New York (1974), p. 7.
11. R. A. Berner, Am. J. Sci., 275, 88 (1975).
12. J. D. Hem, personal communication.
13. G. F. Lee, personal communication.
14. F. Morel and J. Morgan, Environ. Sci. and Tech., 6, 58 (1972).
15. C. E. Weaver and L. D. Pollard, Developments in Sedimentology, 15, The Chemistry of Clay Minerals, Elsevier, New York (1973), Ch. 1.
16. R. L. Malcolm and V. C. Kennedy, J. Water Poll. Cont. Fed., 42, R153 (1970).
17. S. B. Hendricks, Ind. Eng. Chem., 37, 625 (1945).
18. A. F. Frederickson, Bull. Geol. Soc. Am., 62, 221 (1951).
19. W. Stumm and J. J. Morgan, Aquatic Chemistry, Wiley-Interscience, New York (1970), p. 391.
20. G. Millot, Geology of Clays, translated by W. R. Farrand and H. Paquet, Springer Verlag, New York (1970), p. 380.
21. J. D. Hem, C. E. Roberson, C. J. Lind, and W. L. Polzer, Chemistry of Aluminum in Natural Waters, Chemical Interactions of Aluminum with Aqueous Silica at 25° C, U.S. Geol. Surv. Water Supply Paper 1827E (1973).
22. H. Harder, Mineral Soc. Japan Spec. Paper, 1, 106 (1971).
23. H. Harder, Chemical Geology, 14, 241 (1974).
24. H. van Olphen, An Introduction to Clay Colloid Chemistry, Wiley-Interscience, New York (1963).
25. E. A. Jenne in Trace Inorganics in Water, (R. F. Gould, ed.) Advances in Chemistry Series No. 73, Am. Chem. Soc., Washington, D.C. (1968), pp. 337-338.
26. B. J. Anderson and E. A. Jenne, Soil Science, 109, 163 (1970).
27. J. J. Morgan and W. Stumm, J. Colloid Sci., 19, 347 (1964).

28. J. W. Murray, *J. Colloid Interfac. Sci.*, 46, 357 (1974).
29. R. Dawson and E.K. Duursma, *Neth. J. Sea Res.*, 8, 339 (1974).
30. V.C. Kennedy, *Mineralogy and Cation - Exchange Capacity of Sediments from Selected Streams*, U.S. Geol. Surv. Professional Paper 433-D (1965).
31. L. Kamp-Nielsen, *Arch. Hydrobiology*, 73, 218 (1974).
32. B. J. Presley, Y. Kolodny, A. Nissenbaum and I. R. Kaplan, *Geochemica et Cosmochimica Acta*, 36, 1073 (1972).
33. R. Siever, K. C. Beck and R. A. Berner, *J. Geology*, 73, 39 (1965).
34. F. T. Manheim, *Clays and Clay Minerals*, 22, 337 (1974).
35. D. Carroll and H. C. Starky, *Proc. 7th Nat. Conf. on Clays and Clay Minerals*, Washington, D.C., Oct. 20-23, 1958, p. 80.
36. K. L. Russell, *Geochimica et Cosmochimica Acta*, 34, 893 (1970).
37. F. Cerrai, M. G. Mezzandri and C. Triulzi, *Energ. Nucl.*, 16, 378 (1969).
38. Y. Nagaya and M. Saiki, *J. Radiat. Res.*, 8, 37 (1967).
39. Yu. A. Kokotov, R. F. Popova and A. P. Urbanyuk, *Radiokhimiya*, 2, 199 (1961).
40. V. M. Prokhorov, *Radiokhimiya*, 11, 317 (1969).
41. A. Lerman and T. A. Lietzke, *Limnol. Oceanogr.*, 17, 497 (1975).
42. W. A. Deer, R. A. Howie and J. Zussman, *An Introduction to the Rock-Forming Minerals*, John Wiley and Sons, New York (1966).
43. R. E. Grim, *Clay Mineralogy*, 2nd ed., McGraw-Hill, New York (1968), p. 190.
44. Reference 15., p. 4.
45. Reference 15., p. 107.
46. Reference 43.
47. C. E. Marshall, *The Physical Chemistry and Mineralogy of Soil*, John Wiley and Sons, New York (1964).
48. R. K. Iler, *The Colloid Chemistry of Silica and Silicates*, Cornell University Press, Ithaca, New York (1955).
49. Reference 42, p. 225.
50. Reference 15, p. 69.
51. I. Barshad, *Amer. Mineral*, 33, 655 (1948).
52. I. Barshad, *Soil Sci.*, 77, 463 (1954).
53. G. Stanford and W. H. Pierre, *Soil Sci. Soc. Am. Proc.*, 11, 155 (1946).
54. F. H. LeRoux and N. T. Coleman, *Soil Sci. Soc. Am. Proc.*, 27, 619 (1963).

55. F. H. LeRoux, J. G. Cady and N. T. Coleman, Soil Sci. Soc. Am. Proc., 27, 534 (1963).
56. G. R. Frysinger and H. C. Thomas, J. Phys. Chem., 64, 224 (1960).
57. R. K. Schultz, R. Overstreet and I. Barshad, Soil Sci., 89, 16 (1960).
58. G. Eisenman, Biophys. J., 2, 259 (1962).
59. H. Sherry, Ion Exchange (J. A. Marinsky, ed.), Vol. 2, Marcel Dekker, New York (1968).
60. R. M. Barrer and J. Klinowski, J. Chem. Soc. Faraday Trans. I, 70, 2362 (1974).
61. G. E. Likens and F. H. Bormann, Science, 184, 1176 (1974).
62. E. K. Duursma and M. G. Gross, "Marine Sediments and Radioactivity," in Radioactivity in the Marine Environment, U.S. Nat. Acad. Sci., Washington, D.C. (1971), pp. 147-160.
63. Reference 2, pp. 193-194.
64. Reference 2, p. 135.
65. J. V. Lagerwerff and G. H. Bolt, Soil Science, 87, 217 (1959).
66. J. V. Lagerwerff and D. L. Brower, Soil Sci. Soc. Am., 36, 734 (1972).
67. J. V. Lagerwerff and D. L. Brower, Soil Sci. Soc. Am., 37, 11 (1973).
68. W. J. Argesinger, A. W. Davidson, and D. D. Bonner, Trans. Kansas Acad. Sci., 53, 404 (1950).
69. E. Ekedahl, E. Högfeldt, and L. G. Sillén, Acta Chem. Scand., 4, 556 (1950).
70. E. Högfeldt, E. Ekedahl, and L. G. Sillén, Acta Chem. Scand., 4, 828 (1950).
71. J. A. Faucher and H. C. Thomas, J. Chem. Phys., 22, 258 (1954).
72. C. N. Merriam and H. C. Thomas, J. Chem. Phys., 24, 993 (1956).
73. R. J. Lewis and H. C. Thomas, J. Phys. Chem., 67, 1781 (1963).
74. E. H. Cruickshank and P. Meares, Trans. Faraday Soc., 53, 1289 (1957).
75. Reference 5, pp. 155-156.
76. R. A. Robinson and R. H. Stokes, Electrolyte Solutions, 2nd ed. (revised) Butterworths, London (1970), Ch. 15, pp. 432-454.
77. V. M. Rutkovskiy, Oceanology, 13, 211 (1973).
78. Reference 6, p. 26.
79. R. M. Barrer and J. Klinowski, J. Chem. Soc., Faraday Trans. I, 70, 2080 (1974).

80. A. Wild and J. Keay, *J. Soil Sci.*, 15, 135 (1964).
81. E. Wilhelm and R. Wey, *Bull. Serv. Carte. Geol. Alsace Lorraine*, 14, 149 (1961).
82. C. B. Amphlett and L. A. McDonald, *J. Inorg. Nucl. Chem.*, 2, 403 (1956).
83. C. B. Amphlett and L.A. McDonald, *J. Inorg. Nucl. Chem.*, 6, 145 (1958).
84. J. Deist and O. Talibudeen, *J. Soil Sci.*, 18, 125 (1967).
85. R. M. Barrer, L. V. C. Rees and M. Shamsuzzoha, *J. Inorg. Nuclear Chem.*, 28, 629 (1966).
86. H. S. Sherry, *J. Phys. Chem.*, 70, 1158 (1966).
87. H. S. Sherry, *J. Phys. Chem.*, 72, 4086 (1968).
88. R. M. Barrer, J. Klinowski, and H. S. Sherry, *J. Chem. Soc., Faraday Trans. II*, 69, 1669 (1973).
89. M. A. Tamers and H. C. Thomas, *J. Phys. Chem.*, 64, 29 (1960).
90. A. P. Ferris and B. Jepson, *J. Colloid Interfac. Sci.*, 51, 245 (1975).
91. G. R. Frysinger and H. C. Thomas, *Soil Sci.*, 91, 400 (1961).
92. J. S. Wahlberg, J. H. Baker, R. W. Vernon and R. S. Dewer, *U.S. Geol. Surv. Bull. 1140-C* (1965).
93. J. S. Wahlberg and M. J. Fishman, *U.S. Geol. Surv. Bull. 1140-A* (1962).
94. E. A. Jenne and V. C. Kennedy, unpublished data.

Chapter 5

THE THERMAL REGENERATION OF ION-EXCHANGE RESINS

B. A. Bolto

and

D. E. Weiss

CSIRO, Division of Chemical Technology,
South Melbourne, Australia

I.	INTRODUCTION	222
II.	RESIN EQUILIBRIA	224
	A. Equilibrium Characteristics of Individual Resins	225
	B. Structural Factors Influencing Titration Curve Shape	227
	1. Amine resins	227
	2. Carboxylic acid resins	231
	C. Ionic Interactions Influencing Titration Curve Shape	233
	D. Ion Pairing	234
	E. Titration Curve Overlap	237
	1. Selection of the optimum resin pair	237
	2. Selection of the optimum operating conditions	240
	F. Equilibrium Diagrams	242
	1. Derived from titration curves	242
	2. Optimum pH and resin ratio	246
	3. Effect of salt concentration	247
	4. Measurements on mixed resins	250
III.	RESIN KINETICS	253
	A. The Rate Problem	253
	B. Variation of Physical Structure	255
	1. Highly porous resins	255
	2. Amphoteric resins	255
	3. Snake-cage systems	256
	4. Micro particles	256
	5. Composite resins	260
	6. "No-matrix" resins	265

IV.	OPERATION OF THE PROCESS	269
	A. Single Stage Process	269
	B. Multistage Process	271
V.	ENGINEERING ASPECTS	274
	A. Fixed bed pilot plants	275
	B. Continuous contactors	279
VI.	ECONOMIC CONSIDERATIONS	281
	A. Energy Requirement	281
	B. Capital and Operating Costs	281
	C. Pretreatment Requirements	283
VII.	FUTURE DEVELOPMENTS	285
	ACKNOWLEDGMENTS	286
	REFERENCES	286

I. INTRODUCTION

The thermal regeneration of exhausted ion-exchange resins has been the aim of a number of research groups [1-4]. The use of heat rather than chemicals for this purpose provides an obvious economic advantage, in terms of both cost and energy requirements. Furthermore, there is a greatly reduced effluent disposal problem, with no excess of reagents for disposal, but merely the actual amount of salt adsorbed in the service cycle. The variation with temperature of the salt loading of resin systems containing anion- and cation-exchange sites, the concept basic to such operation, has been known for some time. One of the earliest, relevant observations was that of Hatch, Dillon, and Smith [5] in their pioneering work on snake-cage resins, when they noted that the adsorption of salt from a hot solution was somewhat less than that from cold solutions. Rose and co-workers [1] proposed a thermal regeneration process which utilized such resins, and there are several speculative patents centered on other amphoteric systems [3,4]. However, none of these workers realized that the nature of the anion- and cation-exchange sites is critical. Strong electrolyte groups were present in all of the systems, although Bloch [4] considered an amphoteric resin in which weak electrolyte groups were closely spaced, leading to zwitterionic structures. The thermal regeneration effects in all these early attempts were either miniscule or entirely hypothetical.

Our own work on novel methods for the regeneration of ion-exchange systems in the late 1950's centered on the electrical regeneration of carbon electrodes containing ion-exchange sites [6]; change in the acidic and basic strengths of cation- and anion- exchange sites on the carbon surface was sought by applying a potential. This concept, deduced from our study of the literature on active transport in biological systems, led to studies in the field of semi-conducting organic polymers which included the synthesis and measurement of the electrical properties of these materials [7,8]. The principle of electrical regeneration was shown to be sound, but space charge effects were observed which made it very improbable that high exchange capacities could be achieved in these systems when they contained a large number of sites [9]. Study of the dependence of the concentration of current carriers on temperature influenced the eventual direction of our research efforts.

In the biological realm, Klotz and Stryker [10] had shown that the basicity of a dyestuff attached to a protein changes markedly when the protein is denatured by urea. It seemed likely that a similar result could be achieved with heat denaturation. We attempted to introduce thermally sensitive, rather complicated structures into acidic and basic polymers, but it was probably the observation of a marked thermal dependance in the chromatography of proteins on DEAE cellulose [11] that encouraged us to look at less complex systems. Indeed, simple amines show a great variation of k_a with temperature [12]. A study of weak electrolyte resins was therefore undertaken. From this work in CSIRO emerged the invention of ion-exchange resins with equilibrium properties that show a considerable variation with temperature, and the use of hot water rather than chemicals for the regeneration of the salt-loaded resins-- the co-called 'Sirotherm' resins* [13].

---

* Sirotherm is now an ICI Australia Limited trade mark for thermally regenerable ion-exchange resins.

II. RESIN EQUILIBRIA

The essence of the thermal regeneration process is the discovery that a mixture of weakly basic and weakly acidic resins with certain specific structural characteristics [14] can be used to adsorb significant quantities of salt at ambient temperature, and made to release the salt by heating to 80-90°C. The adsorption step involves the transfer of protons from carboxylic acid groups to amino groups to form the cation- and anion- exchange sites:

$$\overline{R'CO_2H} + \overline{R''NR_2} + Na^+ + Cl^- \rightleftharpoons \overline{R'CO_2^-Na^+} + \overline{R''NR_2H^+Cl^-} \quad (1)$$

The equilibrium is temperature sensitive, with both types of groups showing weaker electrolyte behavior on heating, so that the effect is reversed and salt is desorbed to the surrounding solution. For example, a water containing 1000 mg/l of sodium chloride can be treated at ambient temperature in a fixed-bed column to bring the salt content down to the potable limit of 500 mg/l, with a yield of 85%. The remaining 15% of the feed water is used as the regenerant at 90°C, resulting in an effluent of an average concentration of 3800 mg/l. Low salinity waters containing up to 3000 mg/l of salts can be partially desalted in this manner, in a process which is competitive in a number of situations with reverse osmosis, electrodialysis, and conventional ion exchange [15].

The main driving force leading to the shift in the equilibrium is the large increase in the ionization of water which occurs on heating: when the temperature is raised from 25° to 85°C there is a 30-fold increase in the concentration of protons and hydroxyl ions. Hence the acidic and basic regenerants are provided by the hot water, rather than by the addition of two separate reagents. A simplified model of the equilibrium, introduced for the sake of convenience, considers the dissociation of water as the only temperature dependent variable, keeping k_a and k_b for the acidic and basic resins constant [16,17]. The thermal variation of k_a for the basic resin, the parameter measured experimentally, is then entirely attributed to k_w, with $k_a = k_w/k_b$. In fact, both the acidic resin k_a and the basic resin k_b change with temperature to some extent.

While the concept of thermal regeneration is delightfully simple, it is found that in practice the resin equilibrium is extremely complex, being influenced by the detailed resin structure, the acidity and basicity of the groups, the ratio of acidic to basic groups, the resin affinities, the pH, and the ionic strength, as well as by the temperature.

A. Equilibrium Characteristics of Individual Resins

The complexity of the situation was realized at the beginning, and in order to facilitate the selection of a suitable combination of resin types, measurements of the equilibrium properties of the separate resins were made in the form of titration curves. Examination of the combined systems was avoided to preclude the proliferation of experiments that such an approach would have necessitated.

The thermally regenerable performance of the mixture of commercial basic and acidic resins used in our initial studies, De-Acidite M [*] and Amberlite IRC-50[†], can be deduced from the titration curves of the individual resins as shown in Figure 1. When plotted on the same graph, the intersections of the two curves gives the degree of resin loading for the salt concentration and temperature conditions of the titrations. By determining the curves at two different temperatures, corresponding to the adsorption and regeneration conditions, the capacity utilization of the resins in a thermally regenerable system can be predicted. In 1760 mg/l salt solution, the utilization of about 8% (AB) of the resins' total capacity, is expected for this resin pair when present in equimolar proportions.

---

[*] De-Acidite resins are products of the Permutit Co. Ltd., London.
[†] Amberlite resins are products of the Rohm and Haas Co., Philadelphia.

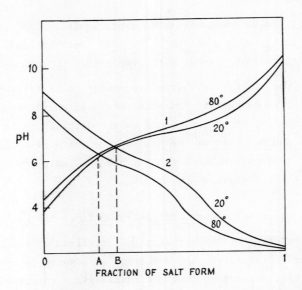

FIG. 1. Titration curves of (1) weakly acidic (Amberlite IRC-50) and (2) weakly basic (De-Acidite M) resins in 1760 mg/l NaCl solution.

The discovery with this approach, that capacity utilization of resin mixtures would be markedly enhanced by improving the buffering capacity of the resins, was vital to the ultimate success of the program. However, a rapid change in pH with degree of neutralization is the normal titration behavior of weakly acidic and weakly basic resins. Polyelectrolyte theory correctly attributes this apparent decrease in the strength of the repeating functional unit as titration proceeds, to the build up of charge on the polymer network (Figure 1). To increase the buffering capacity of a resin (i.e. lower the slope of its titration curve) resins which exhibit extremely flat, plateau-shaped curves were developed. Homofunctionality and other structural requirements in these resin products, described below, needed to be managed, for the successful accomplishment of this objective.

B. Structural Factors Influencing Titration Curve Shape

1. Amine Resins

Resins in which all the basicity arises from tertiary amino groups, with both the resin backbone and the substituent groups on the nitrogen being non-polar in character, yield plateau-type curves on titration. So do crosslinked polyvinylbenzyldialkylamine resins (I) and their cyclic analogues (II) [18]. The curve for I, R=ethyl, is represented in Figure 2 by line 1.

Another suitable tertiary amine resin is that obtained from the polymerization of triallylamine (III) under conditions which avoid

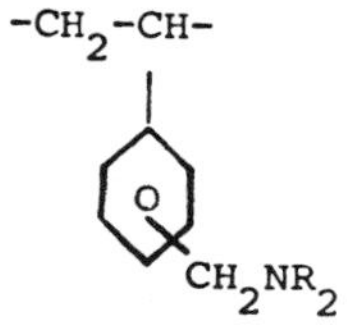

(I) R= $-CH_3$, $-C_2H_5$, $-nC_3H_7$, $-isoC_3H_7$, $-nC_4H_9$, or $-isoC_4H_9$

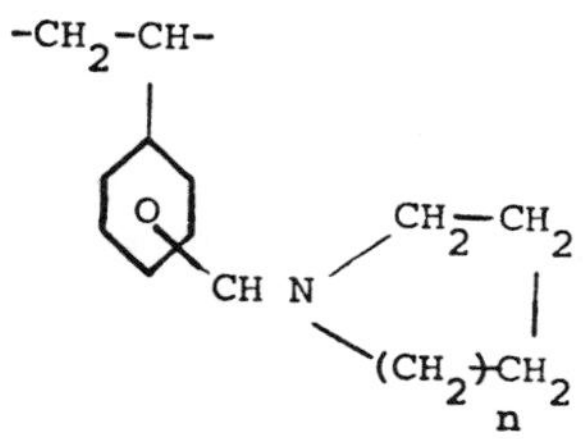

(II) n = 1 or 2

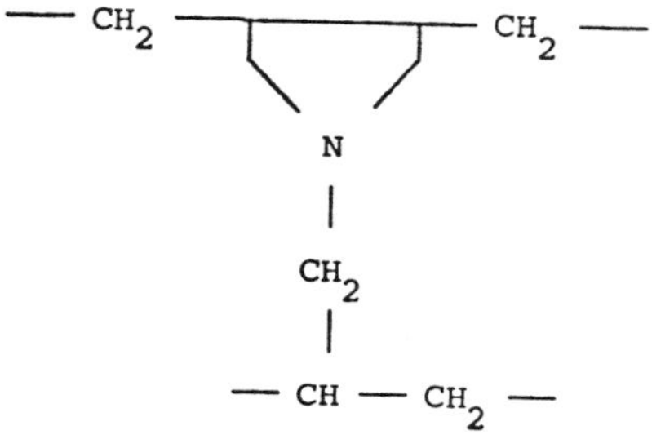

(III)

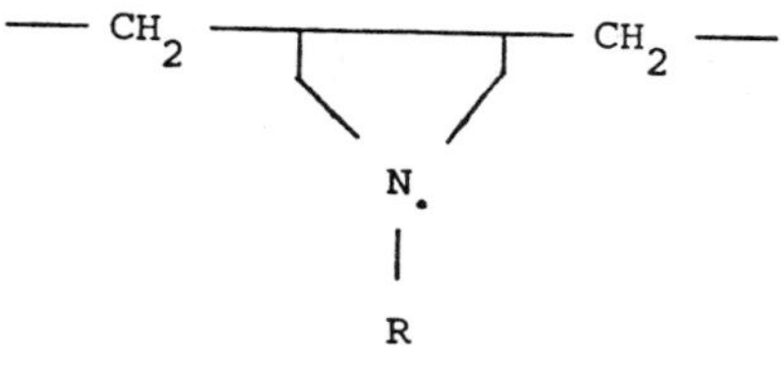

(IV) R= benzyl, $-CH_3$, $-C_2H_5$, $-nC_3H_7$, $-nC_4H_9$

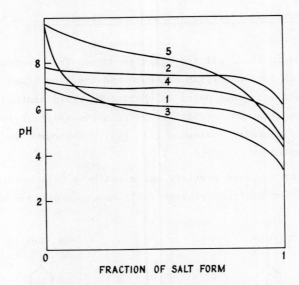

FIG. 2. Titration curves of secondary and tertiary amine resins at 20°C in 1760 mg/l NaCl solution.
1. Crosslinked polyvinylbenzyldiethylamine
2. Polytriallylamine
3. Crosslinked polyvinylbenzyldiethanolamine
4. Crosslinked poly (t-butylaminoethyl methacrylate)
5. Linear diethylaminoethyl cellulose

the incorporation of initiator residues [19,20]. The titration behavior of this resin is also shown (line 2) in Figure 2.

Similar results are obtained with N-substituted polydiallylamines (IV), crosslinked with bis-N-diallylamines, where the substituent is benzyl [21] or alkyl [22].

The improvement in thermal regeneration efficiency when polytriallylamine and poly(acrylic acid) resins are combined is shown in Figure 3, where a much greater proportion, the amount represented by CD or about 40% of the active sites, is available. This result

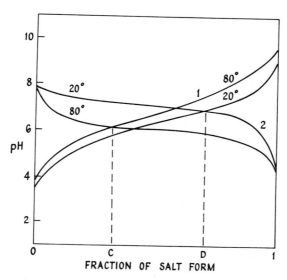

FIG. 3. Titration curves of (1) crosslinked poly(acrylic acid) and (2) polytriallylamine resins in 1760 mg/l NaCl solution.

arises not only from the more advantageous geometry of the overlapping curves, but also from the greater thermal variation of basicity observed with the plateau-producing resins [18]. The major difference between the plateau-producing amine resins and the De-Acidite M resin of Figure 1 is that the De-Acidite M apparently contains a mixture of amino groups of various types (primary, secondary and tertiary) whereas the others are homofunctional.

The importance of homofunctionality as a requirement for introducing buffering capacity into resins can be neatly demonstrated with a polyvinylbenzyldiethylamine resin in which 26% of the nitrogens are in the strongly basic quaternary form, because of the presence of dibenzyldiethylammonium groups. When the free base/OH form is titrated, a sloping curve results (Figure 4). If most of the

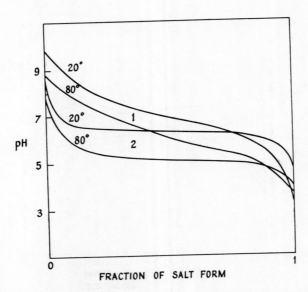

FIG. 4. Titration curves of crosslinked polyvinylbenzyldiethylamine resins in 1760 mg/l NaCl solution.
1. Resin containing 26% of basic groups in the quaternary ammonium form.
2. Resin containing 5.5% quaternary ammonium groups.

ammonium groups are dealkylated by treatment with boiling alkali, to form tertiary amino groups, the resin becomes more homofunctional, with only 5.5% of the nitrogens being left in the quaternary form, and a plateau-shaped titration curve results (also shown in Figure 4).

Secondary amine resins of the polyvinylbenzylalkylamine type contain tertiary dibenzylalkylamine centers as well as the desired secondary amino groups, so that products with less buffering capacity are obtained [18]. However, a strictly homofunctional secondary amine resin (V) can be produced from N-<u>tert</u>.-butylaminoethyl methacrylate and crosslinking agent.

Its titration curve is shown as line 4 in Figure 2 and, like the polyvinylbenzyldiethylamine resin (line 1), exhibits a pronounced plateau [23].

If polar groups are introduced into the resin, as in I where R

[Structures (V), (VI), (VII), (VIII) shown at top of page]

is -CH_2CH_2OH, the titration curve slopes (line 3 in Figure 2). Linear diethylaminoethyl cellulose also exhibits a steep curve (line 5). A similar result has been obtained for a resin made by aminating polyepichlorhydrin with piperidine (VI) and using a diamine as the crosslinking agent [23].

Several homofunctional primary amine resins have been synthesized, including a crosslinked polyvinylamine (VII), a crosslinked polyvinylbenzylamine (VIII), and linear aminoethyl cellulose.

On titration, a sloping curve is obtained for all three, as shown in Figure 5.

Thus primary amine resins, even though completely homofunctional, do not exhibit flat titration curves and the high buffering capacity obtained with a homofunctional tertiary amine resin can be removed by the introduction of polar groups. The above undoubtedly applies to homofunctional secondary amine resins as well.

2. Carboxylic Acid Resins

Crosslinked poly(acrylic acid) resins yield sloping curves on titration (Figure 6, line 1), even though the resin is homofunctional [24]. The shape of the curve is analogous to that of the primary amine resins (Figure 5). The introduction of groups less polar than the carboxylic acid groups flattens the titration curve, but the effect is not nearly as pronounced as with the amine resins. For example, a crosslinked 1:2.6 copolymer of methacrylic acid and methyl methacrylate and a crosslinked 2.3:1 copolymer of acrylic acid and vinyl isobutyl ether both give curves which are flatter than that of the crosslinked poly(acrylic acid) [23]. The curves are represented in Figure 6 by lines 2 and 3.

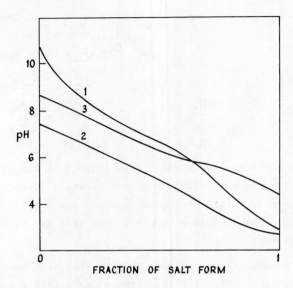

FIG. 5. Titration curves of primary amine resins at 20°C in 1760 mg/l NaCl solution.
1. Crosslinked polyvinylamine
2. Crosslinked polyvinylbenzylamine
3. Linear aminoethyl cellulose

On the other hand, the introduction of vinyl alcohol residues as in a crosslinked hydrolyzed 1:1 copolymer of methyl acrylate and vinyl acetate, yields a steeper curve (Figure 6 line 4); with the high concentration of polar hydroxyl groups in this resin [23], such reduction in buffering capacity is expected.

The importance of plateau character is evident from Figure 3. If an acidic resin could be prepared which yields a flatter titration curve than the poly(acrylic acid) species, but without a significant loss of capacity and lowering of acidity, a much greater practical utilization of resin capacity would be possible. This is shown by comparing EF in Figure 7, the effective utilization of resin capacity estimated for a hypothetical acidic resin of high buffering performance, used in conjunction with a basic resin such as polytriallylamine, with AB in Figure 1, and CD in Figure 3. So far, all attempts to prepare such a resin, including the use of snake-cage structures and resins based on thioglycollic acid [24], have proved unsuccessful.

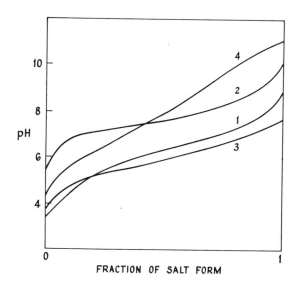

FIG. 6. Titration curves of weakly acidic resins at 20°C in 1760 mg/l NaCl solution.
1. Crosslinked poly(acrylic acid)
2. Crosslinked 1:2.6 copolymer of methacrylic acid and methyl methacrylate
3. Crosslinked 2.3:1 copolymer of acrylic acid and vinyl isobutyl ether
4. Hydrolyzed crosslinked 1:1 copolymer of methyl acrylate and vinyl acetate

C. Ionic Interactions Influencing Titration Curve Shape

Gustafson and Lurio [25] have described the potentiometric curves resulting from titrating a crosslinked poly(methacrylic acid) resin, suspended in a 1.0 M solution of sodium nitrate, with sodium, calcium, nickel, zinc, cupric and ferric hydroxides. They are displaced to lower pH values, and also tend to be less steep, as the interaction between the metal and the resin increases.

The potentiometric titration data obtained during neutralization of a poly(acrylic acid) resin, Zeo-Karb 226[*], with sodium, calcium and lanthanum hydroxides in the presence of their respective

---

[*] A product of the Permutit Co. Ltd., London.

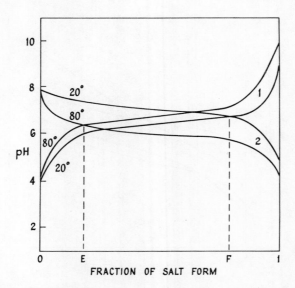

FIG. 7. Titration curves of (1) hypothetical plateau type acidic and (2) polytriallylamine resins in 1760 mg/l NaCl solution.

salts at an ionic strength of unity [23] are shown in Figure 8. The curves are similarly displaced to lower pH values with increasing valency of the metal, due to the increasingly strong interaction of the metal ion with the resin; in addition the curves become flatter with increasing valency and, in the case of the trivalent ion, the initial upturn of the titration curve observed in the sodium hydroxide curve is absent.

Thus it appears that since reducing the polarity of a resin results in the flattening of its titration curve the interaction of multivalent ions with the charged groups attached to the matrix of the resin must result in a lowering of its charge density through covalent binding of the metal.

D. Ion Pairing

Theoretical studies of the titration of a polymeric acid RH with a base BOH usually assume the presence in the polymer of only the two polymeric species RH and R^-, arising from the dissociation of the

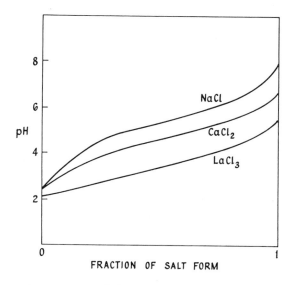

FIG. 8. Titration curves of Zeo-Karb 226 in the presence of various background electrolytes in solutions of unit ionic strength.

acid, since the cation B^+ is assumed to be present in a diffuse ionic atmosphere within the resin phase. Theory [26] then predicts the sloping titration curves observed with the crosslinked poly (acrylic acid) and polyvinylamine resins.

Hamann and Johnson [27] have shown that if a second equilibrium exists between the dissociated salt R^-B^+ and a second species RB, in which B^+ is more strongly bound than in R^-B^+, the titration curve becomes flatter as the equilibrium favors the formation of RB in the reaction $R^- + B^+ \rightleftharpoons RB$. The rapid rise in pH of the initial portion of the titration curves is also reduced, and eventually disappears, as the equilibrium increasingly favors the formation of RB.

It is therefore possible to account for the observed flattening of the titration curves in some weakly basic resins by assuming that the counter ions are chemically (covalently) bound; in other homofunctional resins which exhibit sloping titration curves electrostatic interaction in a diffuse ionic atmosphere prevails.

Titration curves characterized by an extended plateau are observed only in the least polar structures studied. The introduction of alkyl substituents on the amino nitrogen is necessary to produce this titration property. The introduction of polar groups into the hydrocarbon backbone of the polymer, or into the N-alkyl substituents of the amine resins reduces their buffering capacity.

As has been shown by Hamann and Johnson [27], any factor which increases the strength of the electrostatic interactions between the fixed and the counter ions in a homofunctional resin will give rise to a flatter titration curve. The lower local dielectric constant of the less polar resins in the vicinity of the functional groups, the result of displacing water molecules by hydrocarbon chains, would favor ion-pair formation which could account for the plateau in the titration curves, and also for the fact that the basicities of such amine resins are at least 1 pk_a unit weaker than those of the corresponding simple amines [18]. Ion-pair formation would be more favored in amine than in carboxyl resins since the electrical charge density on the functional group is greater in the former case. The carboxyl group is also more polar than an alkylamine group, so that the dielectric constant in the region of the carboxyl group is likely to be higher; this could account for the acidities of lightly crosslinked poly(acrylic acid) resins being little different from those of the corresponding simple acids [24].

The interaction energy between carboxylate groups and ions increases with the valency of the ions, and, in accordance with the theoretical predictions of Hamann and Johnson, could account for the flatter titration curves which result from increasing the valency of the counter ions. Partial solvation of sodium ions by ether or carbonyl groups within the polymeric backbone may occur in the isobutyl vinyl ether copolymer with acrylic acid, and in the copolymer of methyl methacrylate with methacrylic acid. The solvation of alkali metal ions by ester and amide carbonyl groups has been observed with a number of cyclic peptides such as valinomycin [28]. If such an interaction were to increase the strength of binding of the ions, it could account for the flatter titration curves. Evidently the presence of non-polar alkyl groups in these copolymers is an important requirement, since the copolymer containing acrylic acid and vinyl alcohol residues shows the reverse effect.

More recent studies have shown that the overall dielectric constant of a weak electrolyte ion-exchange resin increases as ionization proceeds and a theoretical analysis suggests that this factor also contributes to titration curve flattening [17].

E. Titration Curve Overlap

As mentioned in Section A, the titration curves obtained from single resins can be used to predict the equilibrium behavior of a mixed bed to provide a rational approach to the problems of resin selection and process design [29]. The titration curve of a resin relates the pH of the solution in equilibrium with an amine or carboxylic acid resin to its fractional conversion to the salt form by expressing the amount of acid or alkali added at the specified point in the titration curve, as a fraction of the total ion-exchange capacity of the resin.

The graphical method for prediction of the adsorption of salt by a pair of acidic and basic resins in salt solution from the overlap of the individual resin titration curves when the mixed bed comprises an equimolar mixture of the resins in the undissociated form without any external pH adjustment, has already been demonstrated for several resin pairs (Figures 1, 3 and 7). The difference between the projections onto the resin composition axis of the points of intersection of the titration curves determined under hot and cold conditions indicates the change in composition of the resins when the resin mixture is heated in the salt solution. This thermally regenerable quantity is called the effective capacity (for example, EF in Figure 7).

1. Selection of the Optimum Resin Pair

The importance of the plateau phenomenon has been demonstrated in the examples cited. Equally important is the matching of the

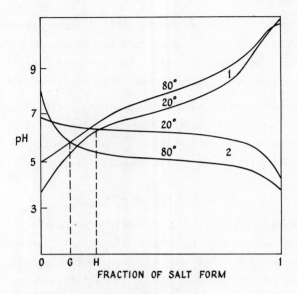

FIG. 9. Titration curves of (1) Amberlite IRC-50 and (2) De-Acidite G in 1760 mg/l NaCl solution.

acidity and basicity of the resins at the relevant salt concentration. This is demonstrated in 1760 mg/l salt solution for De-Acidite G, a polyvinylbenzyldiethylamine resin, matched with Amberlite IRC-50, a poly(methacrylic acid) resin, in one case (Figure 9), and with Zeo-Karb 226, a poly(acrylic acid) resin, in another (Figure 10). The extent of the overlaps (GH in Figure 9, versus IJ in Figure 10) shows that better performance is obtained with the latter system [29].

However, the situation is reversed in very strong salt solution. The resin utilization for the combination containing the poly (methacrylic acid) resin is poor in 1760 mg/l salt solution (GH in (KL in Figure 11). The higher salt concentration is ten times greater than is practical for the process [15]. The extent of overlap changes with salt concentration because of the logarithmic dependence of the plateau levels on ionic strength, both resins behaving as stronger electrolytes as the salt concentration in the external solution is increased [50,18,24].

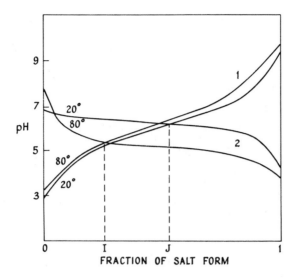

FIG. 10. Titration curves of (1) Zeo-Karb 226 and (2) De-Acidite G in 1760 mg/l NaCl solution. Equilibrium pH 6.2.

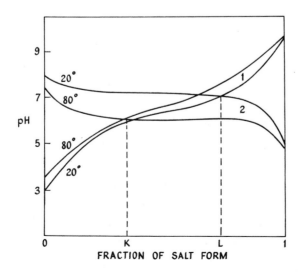

FIG. 11. Titration curves of (1) Amberlite IRC-50 and (2) De-Acidite G in 29,300 mg/l NaCl solution.

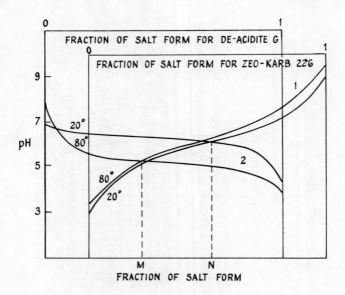

FIG. 12. Titration curves of (1) Zeo-Karb 226 and (2) De-Acidite G in 1760 mg/l NaCl solution, adjusted to correspond to an optimum equilibrium pH of 5.8 at 20°C, using a resin ratio of unity.

There are two major variables in the operation of the process, the importance of which can be gauged by the overlap approach. These are the pH of the system, which can be altered by addition of acid or alkali, and the ratio of acidic to basic sites, or resin ratio. In selecting the best resin pair, care must be taken to ensure that the comparison is made under the conditions optimum for that pair; since these conditions are dependent on resin structure, it follows that each system has its own optimum pH and resin ratio.

2. Selection of the Optimum Operating Conditions

Inspection of the De-Acidite G/Zeo-Karb 226 titration curves (Figure 10), shows that by displacing the curves along the composition axis (Figure 12) a maximum overlap can be obtained. Such displacement of the curves, corresponds to a shift in the equilibrium pH, carried out in this example by the addition of mineral acid to obtain the prescribed pH value at 20°. Often the pH can have a

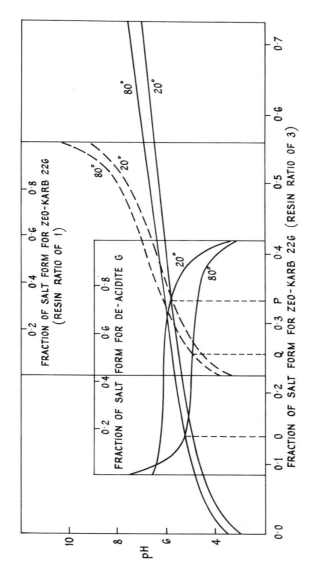

FIG. 13. Titration curves of (1) Zeo-Karb 226 and (2) De-Acidite G in 1760 mg/l NaCl solution, adjusted to correspond to an optimum equilibrium pH of 6.0 at 20°C, using resin ratios of unity and 3.

profound influence on the effective capacity of the mixed resins, particularly when a plateau-type amine resin, in which a large composition change can occur with very little change in pH, is used.

The data presented so far refer only to resin mixtures where the resin ratio is unity. The maximum optimization is not obtainable until both pH and resin ratio are adjusted. For the resin pair under discussion, an increase to a resin ratio of 3 can be depicted by superimposing the titration curves of the acidic resin drawn with the composition axis 3 times as long as that of the basic resin. The relative position of the curves is also moved to select the best pH value for the new system. A dramatic improvement in effective capacity is obtained(cf. compare OP in Figure 13 with MN in Figure 12). The effective capacity is considerably greater at the optimum pH for this resin ratio than with a resin ratio of unity (cf. QP and OP in Figure 13; also compare IJ, MN and QP in Figures 10 12 and 13).

This analysis shows that increasing the resin ratio is tantamount to using an acidic resin exhibiting a flatter titration curve; it represents a practical method for matching the characteristics of a pair of resins not otherwise possible because of the unavailability of acidic resins possessing plateau-type titration curves and high exchange capacities. There is a limit to which the resin ratio may be increased, of course, before the effective capacity per unit weight of the resin mixture begins to decrease.

These considerations show the importance of resin ratio and pH for obtaining the maximum effective capacity of a mixed bed of resins of the weak-electrolyte type [29].

F. Equilibrium Diagrams

1. Derived from Titration Curves

It is customary in the design of two-phase unit operations, such as distillation or solvent extraction, to construct from experimental data an equilibrium curve which shows the composition of one

THE THERMAL REGENERATION OF ION-EXCHANGE RESINS

phase in equilibrium with a second phase, and which forms the basis for calculations of equilibrium conditions in a multistage process. The equivalent diagrams in the present situation are plots of acidic resin ionization versus that of the basic resin at different conditions of temperature and salt concentration. The process can be regarded as one of proton extraction in which one resin phase acts as an extractant for protons from the other resin. Like a vapor-liquid equilibrium diagram, the resin equilibrium diagram is independent of the ratio of the two resin phases, since the pH of a solution in equilibrium with a resin is determined by the composition of the resins.

Such an equilibrium diagram can be constructed from the titration curves of the acidic and basic resins, at the specified temperature and salt concentration, as shown in Figure 14. The equilibrium diagram represents the resin compositions of the two resin phases at the same pH value. Thus points along the curve correspond to different pH values just as points along a vapor-liquid diagram correspond to different temperatures. The titration curves shown are those for De-Acidite G and Zeo-Karb 226 in 1760 mg/l salt solution at ambient temperature. The crosses on the line deduced from the titration curves are obtained from direct experimental determination of the resin compositions of the mixed bed. This excellent agreement provides unambiguous proof for the validity of the method and shows that the two resins behave independently of each other when mixed. Analogous curves derived from titration data at elevated temperatures, are less precise. Nevertheless, the equilibrium diagrams derived at two different temperatures from titration of the individual resins can be used to predict approximate effective capacities for a variety of operating conditions. They are therefore extremely useful in the selection of resins suitable for the thermal regeneration process. The relationship which exists between such equilibrium diagrams and the titration curves of the individual resins is thus of considerable practical importance, since a study of the way in which resin

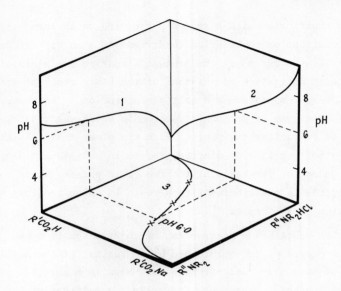

FIG. 14. Construction of the equilibrium diagram for a mixed bed from the titration curves of the individual resins at 20°C in 1760 mg/l NaCl solution.
Three-dimensional representation, viewed as looking into the three faces of an open cube.
1. Titration curve of De-Acidite G
2. Titration curve of Zeo-Karb 226
3. Constructed equilibrium diagram for the Zeo-Karb 226/De-Acidite G mixed bed

structure affects this relationship can be an important guide to the synthesis of improved resin systems [29].

Figure 15 shows a pair of equilibrium curves constructed from the titration curves of De-Acidite G and Zeo-Karb 226 in 1760 mg/l salt solution at 20° and 80°C. The effective capacity of an equimolar mixture of the undissociated resins (that is when there is no external pH adjustment) may be found by drawing a line AB with a slope of unity through the zero point on the resin composition scales. The slope is unity since, for a resin ratio of unity, the change in composition of both resins in the heated and unheated mixed bed must be the same in order to maintain electrical neutrality.

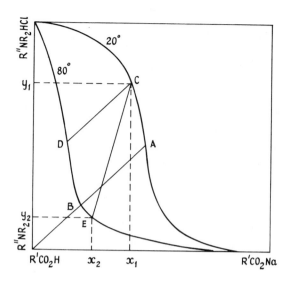

FIG. 15. Equilibrium diagrams for the Zeo-Karb 226/De-Acidite G mixed bed in 1760 mg/l NaCl solution, constructed from the titration curves of the individual resins.

The point of intersection of the line with the equilibrium curve (20°C), A, corresponds to the pH value of the intercept of the titration curves (20°C) when superimposed on a common resin composition scale. The point of intersection of the line with the 80°C equilibrium curve, B, gives the composition of the resins when heated to 80°C. The difference in resin compositions corresponding to points A and B therefore gives the effective capacity. The changes in resin composition which occur when acid or alkali is added to the same resin mixture can be described by parallel lines such as CD, which intersect the 20°C equilibrium curve at the appropriate pH value.

If the resin ratio is changed from unity, the effective capacity can be determined from lines such as CE with a slope corresponding to the resin ratio and intersecting the 20°C equilibrium curve at the desired pH value. For example, if there are three equivalents of acid resin for one of base resin, the change in composition of the acid resin following the adsorption of sodium ions is given by the projection, $x_1 - x_2$, of the line CE, and will correspond to only one-third of the change in composition of the base resin, $y_1 - y_2$, as a result of the adsorption of an equivalent amount of chloride ions.

By such a graphical procedure it is thus possible to calculate from the relevant titration curve data the approximate effective capacities for different resin ratios of the mixed bed in equilibrium with solutions having a variety of pH values.

2. Optimum pH and Resin Ratio

The effect of pH and resin ratio on the effective capacity of mixtures of De-Acidite G and Zeo-Karb 226 is shown in Figure 16, derived as described above from the titration curves. The graph shows that at each pH value there is an optimum resin ratio for

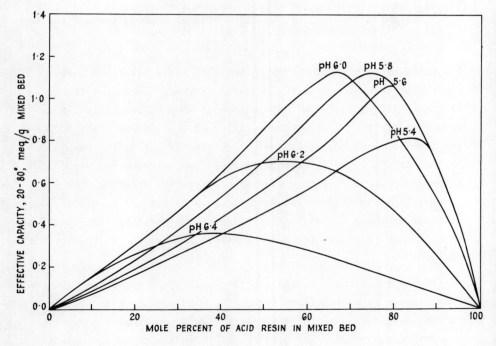

FIG. 16. Predicted effective capacity, 20-80°C, versus composition of the Zeo-Karb 226/De-Acidite G mixed bed in 1760 mg/l NaCl solution at various equilibrium pH levels.

maximum effective capacity, and that the effective capacity varies markedly with changes in the pH level, just as has been predicted with the scanning procedure based on the overlap of the titration curves.

This result may be explained as follows: The pH of the system is determined by the compositions of the acidic and basic resins in the mixed bed. On heating the system, the degree of ionization of both resins is reduced, so that the pH of the system changes until equilibrium is reached again. The greater the ratio of acidic to basic resin, the smaller will be the change in the acidic resin composition on heating, and the smaller the pH change. Consequently, the amount of salt desorbed by heating a mixed bed under equilibrium conditions of constant initial pH and salt concentration will increase with increase in the resin ratio. Since the weight of the mixed bed, for a given amount of basic resin, increases also with increase in the resin ratio, the amount of salt desorbed per unit weight of mixed bed passes through a maximum.

Figure 17 shows the predicted effective capacities at the optimum resin ratio as a function of pH for 1760 mg/l salt solutions and for a variety of amine resins in combination with Zeo-Karb 226. The optimum resin ratio at the maximum effective capacity for each resin pair is given in parentheses in the figure legend. The curves show that resins differ widely in their performance. The effective capacity is critically dependent on pH when the amine resin is of the highly buffering type, but the dependence is less critical for a heterofunctional resin such as De-Acidite M. The superiority of the pH-buffering resins in terms of efficient utilization of the exchange sites in a thermal regeneration process is quite obvious.

3. Effect of Salt Concentration

The shift at ambient temperature of the titration curves of weak electrolyte resins with salt concentration is expressed by the equations derived by Helfferich [50]

Acidic resins: $pk_a = pH + \log[Na^+] - \log[\frac{1}{2}\overline{X}]$

Basic resins: $pk_a = pH - \log[Cl^-] + \log[\frac{1}{2}\overline{X}]$

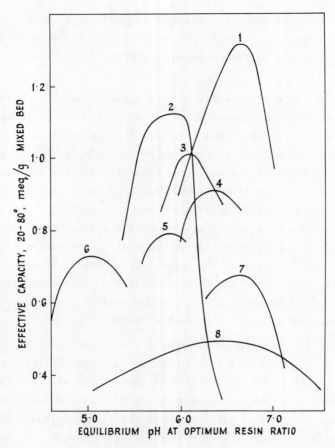

FIG. 17. Predicted effective capacity, 20-80°C, versus equilibrium pH at optimum resin ratio for various mixed beds of amine resins with Zeo-Karb 226 in 1760 mg/l NaCl solution. The optimum resin ratio at the maximum effective capacity is shown in parentheses.

1. Crosslinked polyvinylbenzylethylamine (1.9)
2. De-Acidite G, containing 2.6% quaternary ammonium groups (2.5)
3. Amberlite IRA-93 (2.5)
4. Crosslinked polyvinylbenzyldimethylamine (1.9)
5. As for (2), but with Amberlite IRC-50 as the acidic resin (4.0)
6. Crosslinked polyvinylbenzyldipropylamine (4.0)
7. De-Acidite G, containing 26% quaternary ammonium groups (1.5)
8. De-Acidite M (1.0)

Here the pH refers to that at half neutralization, and $\bar{X}$ is the concentration of ionogenic groups in the resins [18,24].

Since the thermal effect is unaffected by salt concentration [18,24] it is possible to derive titration curves at a variety of salt concentrations from those determined experimentally at one concentration. Using these curves for the individual resins, a family of equilibrium diagrams can be obtained, as shown in Figure 18. These derived data refer to equilibria at 20°C and have been shown to agree well with direct measurements on the mixed resin system [29].

A similar set of equilibrium curves can be deduced for the situation at 80°C, but the data are only approximate. Nevertheless,

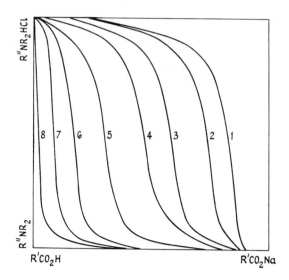

FIG.18. Equilibrium diagrams for the Zeo-Karb 226/De-Acidite G mixed bed at 20°C, constructed from the titration curves of the individual resins in NaCl solutions of the following concentrations.

1. 7500 mg/l
2. 5000 mg/l
3. 3000 mg/l
4. 1760 mg/l
5. 1000 mg/l
6. 500 mg/l
7. 250 mg/l
8. 100 mg/l

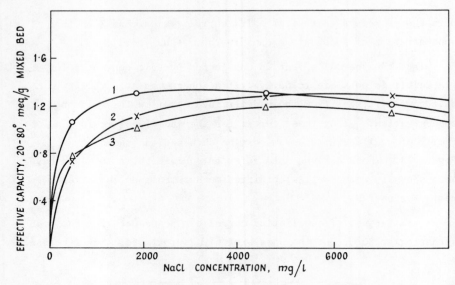

FIG. 19. Predicted effective capacity, 20-80°C, of mixed beds of various amine resins with Zeo-Karb 226 at optimum equilibrium pH and resin ratio, plotted as a function of NaCl concentration. All the data were calculated from the titration curves of the individual resins at 20 and 80°C in 1760 mg/l NaCl solution.
1. Crosslinked polyvinylbenzylethylamine
2. De-Acidite G
3. Amberlite IRA-93

they enable useful predictions to be made of the performance of various resin pairs over a range of salt concentrations (Figure 19).

4. Measurements on Mixed Resins

It is preferable for more detailed studies to obtain the mixed bed equilibria at 80°C by direct measurement. The results for the De-Acidite G/Zeo-Karb 226 mixed bed at a range of salt concentrations are shown in Figure 20. The pronounced effect of temperature on the system is demonstrated by comparing the equilibrium lines with those shown in Figure 18.

Comparison of the performance predictions obtained from the equilibrium diagrams determined experimentally for the mixed resins with those obtained from the titration curves of the individual resins is shown below [29]

Salt Concentration mg/l	Source of Equilibrium Diagrams	Predicted Optimum Data		
		Effective Capacity, 20-80°C meq/g	pH	Resin Ratio (acid/base)
500	Direct measurement	0.8	5.5	3.3
	Titration curves	1.0	5.4	4.0
1760	Direct measurement	1.1	5.9	2.5
	Titration curves	1.5	5.9	2.2

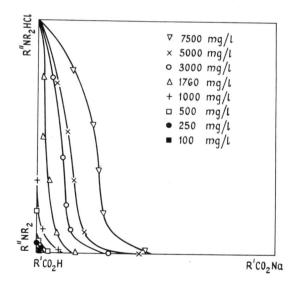

FIG. 20. Equilibrium diagrams for the Zeo-Karb 226/De-Acidite G mixed bed at 80°C, using data determined directly from the mixed beds in NaCl solutions of the concentrations indicated.

The good agreement observed provides strong justification for use of this screening procedure to assess the major differences in performance of various resin pairs. While the predicted effective capacities, expressed as meq of salt taken up at 20° by a unit weight of thermally regenerated mixed bed, are lower than predicted by the more rigorous route, the operating conditions prescribed are essentially the same.

Finally, the predicted effective capacity as a function of pH using the directly measured equilibrium diagrams, is shown as the full line in Figure 21 for the resin ratio of 2.2 estimated to be the optimum for uptake from 1760 mg/l salt solutions. Measurements of effective capacities at a number of pH levels are indicated by crosses in this figure. Good agreement is obtained between exper-

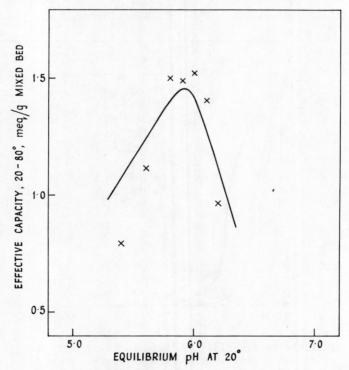

FIG. 21. Effective capacity, 20-80°C, of a Zeo-Karb 226/De-Acidite G mixed bed having a resin ratio of 2.2 versus equilibrium pH in 1760 mg/l NaCl solution.

Full line: prediction from Figure 18 and 20
X: experimental points

iment and prediction to confirm the predictive quality of this approach.

Similar principles apply for systems containing varying concentrations of a single multivalent salt [29]. However the exceedingly complex equilibria which exist in mixtures of various monovalent and divalent electrolytes have so far defied theoretical analysis.

From the extensive studies made of the influence of resin structure on titration curve shape and pk_a, and the subsequent influence of these parameters on the behavior of mixtures of basic and acidic resins of the weak electrolyte type, it has been possible to design resins with improved exchange capacities, with titration plateaus which span a range of more than two pk_a units, and with titration curve shapes controlled to yield equilibrium diagrams which predict practical operation over a wide range of operating conditions.

III. RESIN KINETICS

A major obstacle to the economic operation of a thermal regeneration process is the inordinately slow rate of salt uptake by mixed beds of weak electrolyte resins. An intensive study to remedy this has been undertaken [30-34].

A. The Rate Problem

The diffusion rates of ions being adsorbed by weak electrolyte resins are not greatly different from those obtained with resins of the strong electrolyte type [30,33], yet the overall rates of salt uptake by the two types of mixed beds are greatly different. The weak electrolyte system, as illustrated in Figure 22, is slower by some two orders of magnitude. Of course the strongly dissociated resins are capable of salt splitting and operating independently of one another, whereas the weakly dissociated materials are coupled according to the mixed bed equilibrium (Equation 1). The rate of

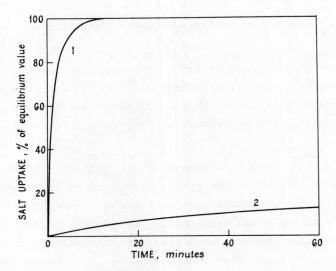

FIG. 22. Rates of salt uptake by mixtures of basic and acidic ion exchangers in 1760 mg/l NaCl solution. The resin ratio is unity, and the particle size 14-52 mesh, BSS.
1. Strong electrolyte resins Amberlite IRA-900/ Amberlite 200 (OH, H forms)
2. Weak electrolyte resins De-Acidite G/Zeo-Karb 226 (free base, free acid forms)

uptake of hydrochloric acid by a weakly basic resin is proportional to the concentration of acid with which it is in contact, at least at concentration levels greater than 10^{-2} M [33]. A similar concentration dependence should apply for alkali uptake by a carboxylic acid resin. However, at the near-neutral pH levels at which the mixed system operates, particle diffusion is most likely rate controlling with the adsorption rate being independent of the reagent concentration [35]. The rate of transfer of protons from the acidic to the basic sites is most compatible with a particle-diffusion controlled mechanism.

A considerable number of approaches have been investigated to speed up the proton-transfer step and hence the rate of adsorption of salt by mixtures of weak electrolyte resins. These have entailed studies of resins of a variety of physical formats.

B. Variation of Physical Structure

1. Highly Porous Resins

The early work to enhance the rate of salt uptake was centered on resins of the gel type, i.e. single-phase beads of very low internal surface area. By using macroporous particles a sizeable rate enhancement was achieved, the increase being about 10-fold for the neutralization of polyvinylbenzyldiakylamine resins with mineral acid. The particular resins used were Amberlite IRA-93 and a resin of the De-Acidite G type, made from macroporous beads [30].

A similar but smaller improvement in the rate of neutralization of poly(acrylic acid) resins by alkali was made by using more porous species. However, the rate of salt uptake by thermally regenerated mixtures of the more porous resins was still too slow by one order of magnitude [31].

2. Amphoteric Resins

Resins containing both acidic and basic groups, incorporated within the same particle on a common polymeric backbone, should adsorb salt very rapidly because of the close proximity of the two types of exchange sites. With this thought in mind, homogeneous amphoteric resins were prepared from macroporous chloromethylated polystyrene beads by reaction with α,ω-amino acids such as piperidine-4-carboxylic acid, γ-aminobutyric acid, or ε-aminocaproic acid.

In all cases rates were further enhanced but extremely poor thermally regenerable capacities were obtained. The best result yielded a thermally regenerable capacity only 5% of what is possible with the De-Acidite G/Zeo-Karb 226 mixed bed. The failure of this approach has been ascribed to the fact that the acidic and basic groups are now too close to achieve independent behavior, so that internal salt structures are produced which exclude the possibility of binding mobile counterions [31]. Even if the sites could be located further apart, the approach also suffers from the inherent disadvantage of a fixed acidic to basic site ratio of unity, while the favored ratio is in the vicinity of 2.

3. Snake-Cage Systems

A series of snake-cage polymers has been synthesized by using modifications of the procedure of Hatch and co-workers [5] to produce linear poly(carboxylic acid) snakes entwined and entrapped in cages of crosslinked amine resins. The products adsorb salt at a suitably rapid rate, but their effective capacities tend to be poor. The highest effective capacity was about 20% of that achieved in mixed bed configurations [31]. This was doubled by using an ester instead of the acidic monomer in the preparation of acidic snakes within macroporous cages [36].

The inevitable conclusion is that the low capacities arise because of the flexibility of the snakes. This results in a high degree of internal neutralization, so that the majority of sites are permanently unavailable for the adsorption of salt.

4. Micro Particles

A reduction in particle size naturally causes an enhancement in rates when the rate-determining step involves diffusion within the particle, the rate being inversely proportional to the square of the particle radius [35]. Such a rate dependence has been confirmed for the reaction of macroporous amine resins with mineral acid [30,33], but peculiar results have been obtained in the case of a resin gel where the rate of neutralization was found to be essentially independent of the particle size, when it was in the 200-400 µm range. This suggests that the chemical reaction is the limiting step. This anomaly has been studied in some detail by Millar and co-workers [37,38].

When the particle size is drastically reduced from the conventional 300-1200 µm to 10-20 µm, the rate of salt uptake by mixed beds of weak electrolyte resins is extremely rapid. There is at least a 100-fold acceleration of the rate even when gel resins are employed [31]. While minute by conventional ion-exchange standards, the particles are of course still macro entities with respect to interactions at the molecular level. Inter particle neutralization effects are therefore absent, so that the capacities observed are the same as those encountered with mixed beds of normal-sized particles.

Physical Format	Particle Size	Effective Capacity, 20-80°C, meq/g	Time to 50% Equilibrium, min.
Mixed bed-gel	Normal	1.1	720
macroporous	Normal	1.0	240
gel	Micro	1.1	2
Snake-cage	Normal	0.4	1

A summary of effective capacity and rate measurements for salt uptake from 1000 mg/l salt solutions is given below for thermally regenerable systems based on polyvinylbenzyldialkylamine and poly(acrylic acid) structures of various physical forms [31,36].

It can be seen that a system which maintains the exchange sites in intimate contact, on a molecular level, as in snake-cage resins, reacts at a superior rate, but is handicapped by a significant loss of capacity. With microparticles such loss in capacity is avoided. However, attempts to exploit the superior rates of micro beads are inevitably thwarted by mechanical handling difficulties: the use of such small particles poses serious problems because of pressure drops and the lack of a suitable method of removing accumulated colloids, since backwashing is not feasible. One answer to this impasse is to prepare microbeads which contain a ferromagnetic component. The magnetized resins are then self flocculating, and settle very quickly, at a rate comparable with normal resins, yet the flocs may be easily sheared by stirring, so that the very rapid reaction rates which prevail because of the micro size of the particles can still be achieved [39]. Magnetic particles therefore combine the rate advantage of very fine resins with the rapid settling properties of much coarser particles.

Magnetic micro beads have been prepared in weakly basic forms, an example is a crosslinked polyethylenimine containing about 8% by volume of γ-iron oxide [40]. Photomicrographs of this resin before and after magnetization are shown in Figures 23 and 24. In Figure 24, the development of chain-like closed ring structures

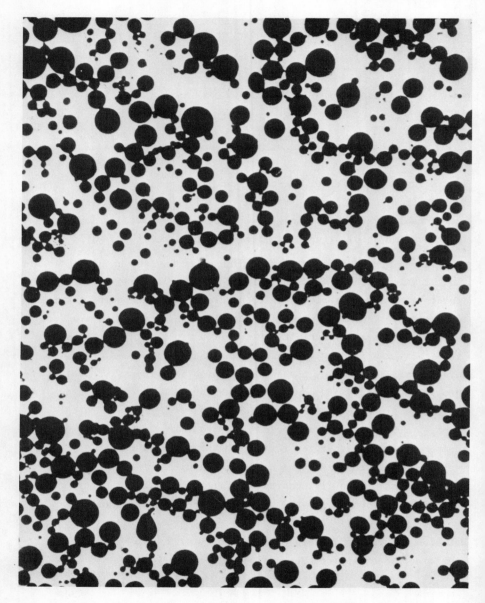

FIG. 23. Crosslinked polyethylenimine beads containing 8% γ-iron oxide, v/v. Bead size 52-120 mesh, BSS. Unmagnetized form.

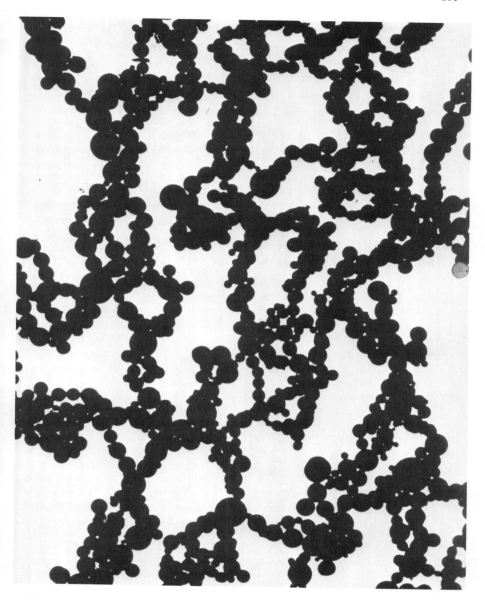

FIG. 24. The magnetic resin particles of Fig. 23 in magnetized form.

making up the floc is shown. A weakly acidic resin with magnetic properties that is based on poly(acrylic acid) has also been prepared [41], as have strong electrolyte types [39].

Because of the unusual properties of the magnetized resins, it has become necessary to initiate a program for the development of completely new contacting procedures. To this end, the dealkalization of hard, alkaline waters by the weakly acidic magnetic resin, regenerated with mineral acid, has been selected for detailed study [41]. Because of the high voidage of the self flocculating particles, their direct transport via peristaltic pumps is possible and a truly continuous counter current operation is feasible for the system. For preliminary studies designed to select the best contacting method a fluidized bed has been built. It is characterized by low resin inventory and great simplicity of design [41,42]. This technique is also being applied to thermally regenerable systems.

5. Composite Resins

In conventional ion-exchange technology, normal-sized particles are preferred. To achieve such utilization of the micro particles the two types of exchanger present in the desired ratio are embedded in a matrix permeable to water and salt [43]. The ensemble is a particle of normal size. These "plum pudding" resins, shown diagrammatically in Figure 25, have been made from crushed commercial resins such as De-Aciditie G (1-10 μm) and Zeo-Karb 226 (5-10 μm), embedded in hydrophilic polymers such as cellulosics, ionically crosslinked polysalts, or poly(vinyl alcohol) crosslinked with a dialdehyde [34]. The particles which result have the standard dimensions of ion-exchange materials, so that completely conventional handling procedures may be used -- aside from the thermal regeneration. Their adsorption rates, corresponding to the rates normally encountered in ion-exchange technology, are at least 100 times faster than a traditional mixed bed of weak electrolyte

THE THERMAL REGENERATION OF ION-EXCHANGE RESINS

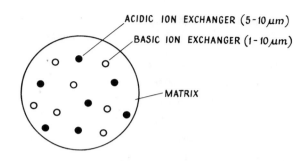

OVERALL SIZE OF COMPOSITE PARTICLE 300-1200 μm

FIG. 25. Diagrammatic representation of a "plum pudding" resin bead.

resins (Figure 26). The rate is quite dependent on the matrix material, being slower when the matrix is ethyl cellulose or a polysalt made from poly(vinylbenzyltrimethylammonium chloride) and sodium polystyrenesulfonate, and fastest with the crosslinked PVA matrix.

That the adsorption rates are influenced by the amount of matrix present is demonstrated by the data below for "plum pudding"

Matrix	Amount of Matrix weight %	Time to 50% equilibrium min
Polysalt	16	5.0
	28	29
	43	35
PVA corsslinked with glutaraldehyde	30	2.3
	40	3.6
	67	20
PVA crosslinked with terephthaldehyde	60	26

resins containing polysalt or crosslinked PVA matrices. The polysalt was prepared from polyelectrolytes with a molecular weight of approximately 4×10^5, and a degree of substitution of 0.6. The

PVA was crosslinked with sufficient dialdehyde to react with 20% of the hydroxyl groups.

In both instances the rates slacken as the matrix content is increased, because there is more matrix for the ions to permeate. This suggests that diffusion of ions through the matrix is the rate-limiting step. Variations in the chain length and charge density of the polyelectrolytes employed in the preparation of polysalt matrices, and in the actual structure of the cationic polyelectrolyte, have remarkably little effect on the adsorption of ions by the micro-ion exchangers within the matrix [34]. Similarly, as shown above, a variation in the nature of the crosslinking agent does not greatly alter the salt-adsorption properties of "plum pudding" resins with a PVA matrix.

For a fixed quantity of a particular type of matrix, the rate varies with the size of the composite bead. Thus in a resin consisting of crushed Amberlite IRA-93 (1-3 μm) and Zeo-Karb 226 (5-10 μm), contained by 40% by weight of crosslinked PVA binder, the half-time of the reaction varies as follows:

420-700 μm beads	1.0 min
180-300	0.4

Experiments in which the speed of agitation of the particles was varied have shown that the rate of uptake remains constant with stirrer speeds ranging from 530 to 1400 rpm, to suggest that diffusion through the static film of solution around the composite resin bead is much faster than diffusion within the bead. The faster reaction with the smaller beads is confirmation that the matrix is the principal diffusional barrier. However, the inverse dependence of the rate on particle size is almost linear, and not to the power of 2 as required by a normal particle-controlled mechanism. That the matrix does not provide the sole diffusional resistance is quite evident from the fact that when only the amine resin component is altered, from De-Acidite G (1-10 μm) to the smaller and inherently faster reacting Amberlite IRA-93 (1-3 μm),

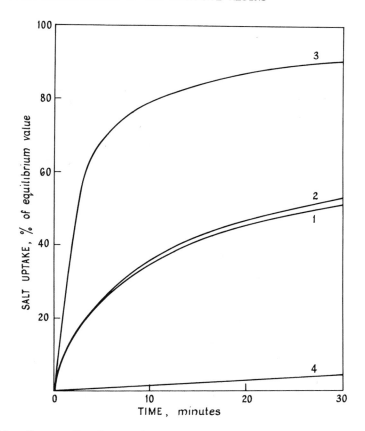

FIG. 26. Rates of salt uptake by thermally regenerated "plum pudding" resins in 1170 mg/l NaCl solution. The resins all contain De-Acidite G/Zeo-Karb 226 particles, have a resin ratio of 2.5, particle size 14-52 mesh, BSS, and contain 30% by weight of the following matrices.

1. Ethyl cellulose
2. Polysalt
3. Crosslinked PVA
4. Normal De-Acidite G/Zeo-Karb 226 mixed bed

the half-time of the reaction is almost halved. The mechanism of adsorption in the "plum pudding" resins is thus quite complex, with both the matrix and the amine resin component supplying barriers to the entry of ions.

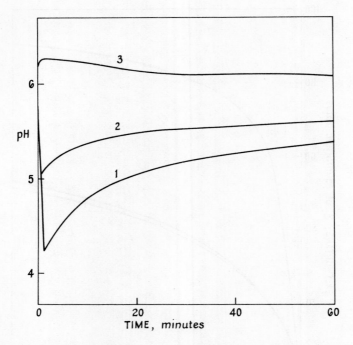

FIG. 27. Variation of solution pH during uptake of salt by thermally regenerated "plum pudding" resins containing 40% by weight of crosslinked PVA as matrix and the following weakly basic ion exchangers in addition to Zeo-Karb 226 (5-10μm).

1. De-Acidite G (10-20μm)
2. De-Acidite G (1-10μm)
3. Amberlite IRA-93 (1-3μm)

Further insight with respect to the mechanism of adsorption in the "plum pudding" resin is available from a study of the pH change with time that occurs during salt adsorption when only the anion resin component of the matrix is varied. That De-Acidite G reacts more slowly than the acidic component in the "plum pudding" resin cited is shown in Figure 27 by the initial release of protons as adsorption of salt by the thermally regenerated composite beads commences. The magnitude of proton release becomes smaller when the size of the De-Acidite G particle is reduced from 10-20 μm to

1-10 μm; the pH imbalance virtually disappears when the gel-type resin is replaced with 1-3 μm particles of macroporous Amberlite IRA-93. With proper matching of the basic and acidic micro-ion exchangers in the matrix to bring them into balance, an adsorption mechanism which embraces the diffusion of ions through the matrix as the rate limiting step is much more probable.

The dilution of the active ion exchangers with an inert matrix naturally means that the ion-exchange capacity of the resin product is correspondingly diminished. For "plum pudding" resins containing 40% matrix by weight, the effective capacity can only be 60% of the level achieved in an undiluted mixed bed. It is the amount of salt adsorbed per unit time that is important, however. In the practical adsorption cycle time of about 30 minutes [15] most of the "plum pudding" resin's thermally regenerable capacity can be utilized while only 4% of the effective capacity of the standard mixed bed (Figure 26) is used; the actual amount of salt removed is increased 13-fold by use of "plum pudding" resins.

With improvements to the basic components employed, "plum pudding" resins can achieve effective capacity performances comparable with the old style mixed bed based on polyvinylbenzyldialkylamine and poly(acrylic acid) resins while the adsorption rates are enhanced by the required 2 orders of magnitude.

The composite resin configuration has the advantage that magnetic forms of the resin can be readily synthesized, by adding $\gamma\text{-}Fe_2O_3$ as a third component to be bound by the matrix. With this modification the use of smaller composite beads to provide an additional enhancement of adsorption rates can be achieved; the half-time of the reaction can be reduced to less than 1 min by this approach.

6. "No-Matrix" Resins

It is apparent from analysis of the results with composite resins that another resin format which could provide desirable capacity and rate performance is a three-dimensional mozaic structure, which can best be pictured as a "plum pudding" resin devoid of matrix, in which the active domains have been grafted to

one another to yield a highly porous, conventional-sized bead.
Such a "no-matrix" resin could have the capacity of a mixed bed,
without the handicap of inert filler, and faster reaction rates
than the three component resins. However, attainment of the full
potential of such a resin would be subject to interference by the
tendency for interaction between the basic and acidic groups which
could lead to the incidence of substantial zones of inactive
material. The simplest preparative route would involve the direct
production of beads in one or two steps from the appropriate
monomers.

Unfortunately, the direct use of acidic and basic monomers is
not fruitful, for the reason cited above. Negligible capacities
are achieved because of internal salt formation between the
oppositely charged sites as was the case with the amphoteric and
snake-cage systems. However, when precautions are taken to
prevent such interaction, useful resins can be produced.

For example, when a homogeneous solution of triallylamine and
methacrylic acid is polymerized at pH 4-5 in the presence of added
anions and cations, particularly multivalent or large organic ions,
to suppress electrostatic interaction between the monomers, crosslinked resins can be produced which have the following properties,
after removal of the added counterions by washing with acid or
alkali [44].

Counterions for Acidic Monomer	Counterions for Basic Monomer	Effective Capacity, 20-80°C, meq/g
Nil	Nil	0.0
Na^+	SO_4^{--}	0.3
Mg^{++}	Cl^-	0.2
Ca^{++}	Cl^-	0.5
Sr^{++}	Cl^-	0.2
Ba^{++}	Cl^-	0.3
Zn^{++}	Cl^-	0.6
Mn^{++}	Cl^-	0.4
Benzyl NMe_3^+	Cl^-	0.5

THE THERMAL REGENERATION OF ION-EXCHANGE RESINS

Likewise, a resin made from triallylamine and acrylic acid in the presence of sodium, chloride, and sulfate ions has an effective capacity of 0.6 meq/g, half of which can be utilized in 2 min. This rate is comparable with the rate achievable with a mixed bed of microparticles. However, the effective capacity is only about a third of that of the analogous mixed bed, indicating that substantial internal salt formation is still occurring.

Better results are possible if a neutral precursor which can be hydrolyzed in the final resin product to provide the acidic groups [45] is employed. Thus if a homogeneous solution of triallylamine hydrochloride and methacrylamide is polymerized, the product after alkaline hydrolysis is a crosslinked resin of effective capacity 1.0 meq/g. The capacity can be raised to 1.2 meq/g by the additional incorporation of diallylamine hydrochloride. About two thirds of the sites can therefore be used effectively. The extraordinary feature of this simplistic approach via the two monomers is that electron micrographs of suitably stained and sectioned beads clearly show the presence of well-ordered domains of about 1 μm in size for the two types of groups (Figure 28).

Resins with an effective capacity of 0.9 meq/g have been prepared by using acrylamide or methyl acrylate in lieu of methacrylamide. It can be advantageous to add another crosslinking agent, such as methylene-bis-acrylamide or ethylene glycol dimethacrylate.

The adsorption half-time in general ranges from 2 to 15 min. for the neutral precursor products, this parameter being dependent on the physical strength of the material with the tougher, more crosslinked resins reacting the slowest. The structures of the resins are not well defined. While block copolymerization must occur to some degree, differential thermal analysis and experiments on linear analogues [46] show that the two homopolymers are certainly present, presumbly in snake-cage or interpenetrating network configurations.

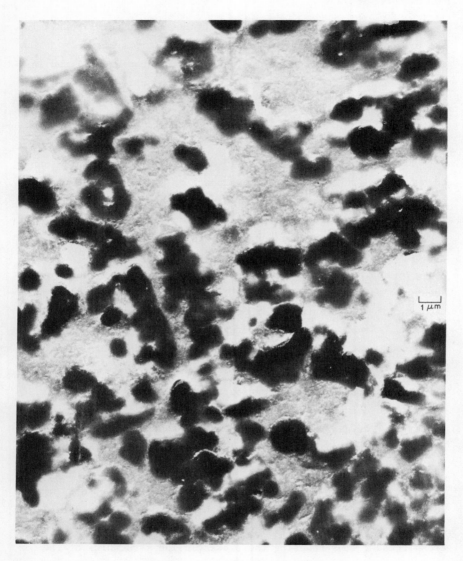

FIG. 28. Electron micrograph of a stained, sectioned no-matrix resin bead made by polymerizing triallylamine hydrochloride and methacrylamide, followed by hydrolysis. The acidic domains are more darkly stained.

There are other modes of resin synthesis applicable to no-matrix resins, which are capable of producing high capacity resins, but

THE THERMAL REGENERATION OF ION-EXCHANGE RESINS

which require more extensive or complicated routes; for example by starting with one of the components in the solid phase [47].

IV. OPERATION OF THE PROCESS

A. Single Stage Process

The simplest form of flowsheet for a thermal regeneration process uses a single fixed bed of resin [48]. A typical fixed-bed concentration and temperature profile is shown in Figure 29. Feed water at 20°C is passed through the bed to yield product water of the desired quality, until the resin is heavily loaded with salt. At this stage the product water quality deteriorates. The resin is then restored to its initial condition by regeneration with feed water, heated to 80°C in this example. In most cases the tail end of the product water is preferred as regenerant. Heat economy is achieved by using the hot effluent to warm the incoming regenerant water.

To conserve water, some of the water displaced when heating and cooling the bed can be recycled. The mode of operation can be concurrent or countercurrent. The total cycle typically occupies one hour and is automatically controlled. The yield of product water from feed water in fixed bed operation is usually between 70 and 85%, depending on the feed water concentration.

The equilibrium properties of the resins are pH dependent, so that the resin must be preloaded with acid or alkali to the optimum pH level. The results listed below for waters containing sodium chloride, or calcium chloride, respectively, show the influence of pH as well as the pH for optimum operation with these two salts.

Resin equilibrium pH (in 1000 mg/l NaCl)	6.9	7.1	7.4	7.6	8.1	8.7
Operating capacity*(meq/g)						
Feed of 500 mg/l NaCl only	0.35	0.50	0.32	0.16	0.05	0.01
Feed of 570 mg/l $CaCl_2$ only	0.14	0.27	0.45	0.56	0.56	0.36

*The operating capacity is the salt adsorbed by a unit quantity of the resin in one cycle once the column has reached a steady state condition.

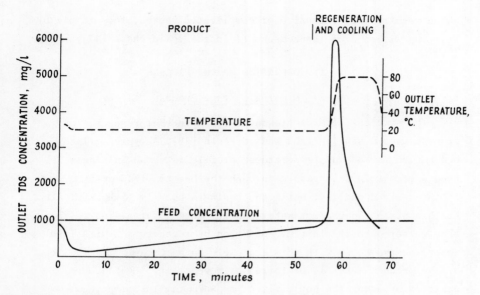

FIG. 29. Typical salt concentration and temperature profiles for a thermally regenerated system treating water containing 1000 mg/l NaCl.

When handling buffered-feed waters, such as those containing bicarbonate, the pH of the feed water must be controlled so that, after passage through the column, there is no net change in resin pH.

Some typical column runs on a water containing only sodium chloride are shown in Table 1, using adsorption-flow rates ranging from 0.18 to 0.80 bed volumes (BV)/min. The adsorption stages were operated to obtain an average product concentration of 500 mg/l from a feed containing 1000 mg/l of dissolved salt. The results [15] show that the adsorption step is to some extent kinetically limited, but very good product yields can be obtained even with flow rates as high as 0.80 BV/min. Some compromise of high yields and product quality has to be made with high-throughout operation. At the elevated temperature of regeneration the kinetics are understandably faster, so that this step in the process is equilibrium limiting, and there is no loss in efficiency whatsoever for regeneration flow rates up to 0.50 BV/min. The performance of the process is also illustrated in Table 1 for feed waters containing lower salt contents of 540 and 300 mg/l, respectively.

Although resins containing carboxylic acid groups are especially selective for alkaline earth cations, the process can be used to treat waters containing magnesium sulfate [49] or calcium chloride [15]. The resin used in Table 1 will quite satisfactorily handle a water containing only calcium chloride, as shown in Table 2. Successful operation is achieved by adjusting the resin to the higher equilibrium pH required for optimum performance with the calcium chloride feed. The results parallel those obtained with sodium chloride waters.

For waters which contain a mixture of cations, but which are predominantly hard, the divalent cations can be selectively removed, with the coions, in a resin equilibrated to the higher pH level; the monovalent cations are weakly adsorbed, and simply leak through the column [15].

However, when a predominantly soft water is treated at pH levels suitable for the removal of monovalent cations, the divalent cations are strongly held at ambient temperature, so that the performance is greatly enhanced by first removing any hardness completely, by conventional ion-exchange softening or softening dealkalization.

No similar trend in anion selectivity is observed, sulfate being adsorbed and regenerated like chloride.

B. Multistage Process

Because of the selectivity of weak electrolyte systems for divalent (hardness) ions, it is not usually the most economic procedure to remove significant amounts of both divalent and monovalent cations in the one step. Instead, such a water is treated best by removing divalent and monovalent cations in separate thermally regenerable columns. The steps used in the treatment of a water of complex analysis would be as follows.

1. Dealkalization

This step removes alkalinity and an equivalent amount of hardness by the conventional ion-exchange process. The step

TABLE 1

LABORATORY COLUMN PERFORMANCE ON WATERS CONTAINING NaCl ONLY
RESIN pH (IN 1000 mg/l NaCl) = 7.2

Feed concentration	(mg/l)	1000	1000	1000	1000	540	300
Adsorption	(at ~ 20°C)						
Flowrate	(BV/min)	0.18	0.48	0.80	0.36	0.38	
Time	(min)	132	43	22	60	90	
Regeneration	(by feed at ~ 85°C)						
Flowrate	(BV/min)	0.32	0.32	0.32	0.36	0.38	
Time	(min)	13	13	13	12	14	
Operating capacity	(meq/ml)	0.20	0.18	0.15	0.14	0.10	
Average product	(mg/l)	500	500	500	180	135	
Yield	(%)	85	83	81	83	86	

TABLE 2

LABORATORY COLUMN PERFORMANCE ON WATERS CONTAINING $CaCl_2$ ONLY
RESIN pH (IN 1000 mg/l NaCl) = 8.1

Feed concentration	(mg/l)	920	470	290	280
Adsorption	(at ~ 20°C)				
Flowrate	(BV/min)	0.27	0.35	0.33	0.59
Time	(min)	50	60	151	85
Regeneration	(by feed at ~ 85°C)				
Flowrate	(BV/min)	0.25	0.35	0.34	0.34
Time	(min)	16	14	24	24
Operating capacity	(meq/ml)	0.12	0.12	0.16	0.13
Average product	(mg/l)	430	170	110	130
Yield	(%)	77	81	86	86

is not essential but is efficient on highly alkaline waters. If dealkalization is omitted, close pH control of the feed water is required.

2. <u>Sirotherm on a Divalent Cycle</u>

Divalent cations, and an equivalent amount of anions, are removed in a thermally regenerable system optimized for that purpose. The monovalent cation concentration is largely unaffected.

3. <u>Softening with a Strong Electrolyte Resin or a Carboxyl Resin in Sodium-Form</u>

The product from this step is virtually free of calcium and magnesium.

4. <u>Sirotherm on a Monovalent Cycle</u>

The final step in reducing the dissolved solids to the required level involves the use of a second thermally regenerable system optimized for that purpose. This step is fairly intolerant to low levels of calcium or magnesium.

It must be emphasized that all four steps would not necessarily be employed for the treatment of a particular water. For instance, in the case of a hard water containing little sodium chloride, the process could be terminated after step 2, yielding a soft potable water. In the case of a predominantly soft water, it would be sufficient to combine steps 1 or 3 with step 4.

The resin employed in step 2 for divalent cation removal would be a more weakly adsorbing system than that employed in step 4 for monovalent cation removal. This objective can be achieved by equilibrating the same resin to different pH values; it may be preferable to use two resins of different chemical structure for this purpose.

V. ENGINEERING ASPECTS

Thermal regeneration research has been oriented towards the development of new resins; consequently most of the engineering work has been aimed at employing conventional ion-exchange technology wherever possible. The engineering principles involved in adsorption plants are simple, versatile and highly developed; they are used in both very large and small scale plants for sand filtration and conventional ion-exchange softening and demineralization. Most of the work discussed here has employed fixed bed units of standard type. Some pilot scale work has been conducted in a Boby continuous unit based on the Asahi principle, and in a simulated continuous countercurrent contactor.

A. Fixed Bed Pilot Plants

Three fixed bed units have been used to test the process on the pilot scale [48]. A unit with outputs of 50 to 100 m^3/day was operated for a total of 4200 hours in Melbourne on clarified, deaerated tap water to which 800 to 3000 mg/l of salts were added. A typical example of column performance is shown in Table 3.

This unit was transported to Perth, Western Australia, and operated by the local Water Board on deep artesian well water. The unit is shown in Figure 30. The only pretreatment required has been the adjustment of the well water pH with small quantities of mineral acid, and cooling. In 5700 hours of operation, the performance of the unit has indicated the high reliability and low supervision requirements desired from a plant of this type. The raw water composition and typical performance data are shown in Table 3. Current indications from the laboratory studies discussed above are that improved performance will be achieved by completely softening the feed water in a pretreatment stage.

A smaller unit of 20 m^3/day has been operated by the Australian Mineral Development Laboratories (AMDEL) in Adelaide for a total of 4500 hours to study performance on synthetic feed waters in the 1000 to 2000 mg/l range. Extensive studies have also been made to determine the pretreatment required for surface waters containing significant turbidity and color. Combinations of the thermal regeneration process with other ion-exchange operations such as dealkalization and softening have been studied. The unit has

TABLE 3

PERFORMANCE DATA FOR FIXED BED PILOT PLANTS

		PERTH	MELBOURNE	ADELAIDE	ADELAIDE
Feed water source		artesian well	clarified mains water dosed with salts	surface water after lime softening, etc.	well water after dealkalization
Feed Water Analysis	TDS mg/l	1220	1600	280	714
	Na mg/l	388	545	58	178
	Ca mg/l	18	38	20	52
	Mg mg/l	8	-	13	26
	Cl mg/l	519	716	110	408
	HCO$_3$ mg/l	201	209	37	-
	SO$_4$ mg/l	39	100	55	48
Product Water	average mg/l	560	673	80	473
	minimum mg/l	280	500	60	-
Effluent	average mg/l	3550	3400	835	1980
	minimum mg/l	4020	6750	1550	-
Adsorption flowrate	BV/min	0.21	0.19	0.24	0.21
Yield	%	79	71	84	84
Regeneration temperature	°C	90	93	82	90
Adsorption temperature	°C	29	25	14	19

	PERTH	MELBOURNE	ADELAIDE	ADELAIDE

THE THERMAL REGENERATION OF ION-EXCHANGE RESINS

FIG. 30. The 0.45 m diameter thermally regenerable column installed at Perth.

also been operated on a well water of high hardness. A typical result is shown in Table 3.

Studies on the pretreatment requirements for surface waters were combined with an evaluation of the process as a roughing operation ahead of a conventional mixed-bed demineralization plant making boiler feed water. A typical result from this trial is also shown in Table 3. Because of the absence of magnesium in the product water it is feasible to use a conventional mixed bed at higher feed water concentrations without impairment of final water quality.

A commercial unit of 600 m^3/day capacity has been constructed by ICI Australia at one of its factories in South Australia

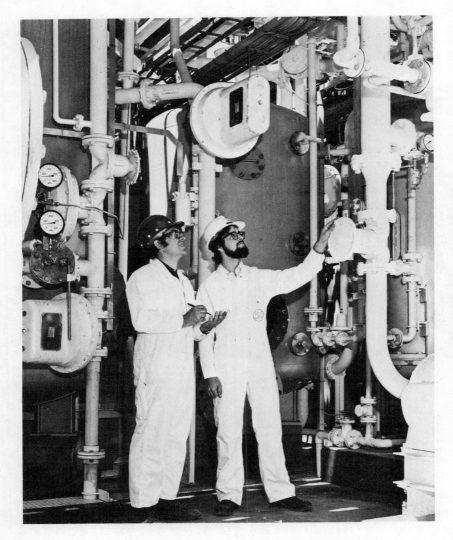

FIG. 31. Commercial installation at Osborne, South Australia, capable of treating 600 m^3/day of saline surface water, using Sirotherm resin in a fixed bed.

(Figure 31). This is an example of the application of the process as a roughing operation in the purification of saline surface waters for industrial usage, and represents the first commercial operation of the process.

B. Continuous Contactors

A continuous or nearly continuous output of product water can be achieved by using multiple fixed beds and staggering the cycles. In the limit, with sufficient fixed beds operating out of phase, the operation closely simulates continuous countercurrent operation. A contactor employing this principle has been operated on the 60 m^3/day scale. The practical problems encountered in using multiple beds have been overcome by combining twenty beds in a single vertical column containing twenty-one distributors and progressively switching the flows through a multiport valve (Figure 32). As shown in Table 4, very high yields of product water (up to 90% of the feed water) can be achieved, so that this method appears very promising.

Commercial moving bed units, such as the Asahi contactor, are also operated to approximate continuous countercurrent operation by intermittently moving batches of resin through a transfer loop.

TABLE 4

PERFORMANCE OF CONTINUOUS CONTACTORS

Type of contactor			Boby Cl, Asahi type	Simulated continuous countercurrent
Product water output		m^3/day	48	54
Feed water	TDS	mg/l	1120	1190
Product water	TDS	mg/l	530	410
Effluent	TDS	mg/l	5970	6000
Yield of product water		%	90	85
Regeneration temperature		°C	85	95

FIG. 32. A fixed bed pilot contactor simulating continuous flow by the interswitching of hot and cold flows through 20 beds.

A 0.3 m diameter Boby CI (Asahi) moving bed contactor was constructed by AMDEL for use in thermal regeneration (Figure 33). The plant has proved to be simple to operate and mechanically reliable. It has shown high yields of product water, good thermal economy and a low rate of resin attrition. Typical performance data are shown in Table 4. However, resin transfer is sluggish since the Sirotherm resin is more deformable than many conventional resins. Modifications to the equipment configuration and the use of more rigid beads will increase the output from continuous contactors.

VI. ECONOMIC CONSIDERATIONS

A. Energy Requirement

When compared with evaporative desalination processes, a thermally regenerable ion-exchange process has a heat requirement low in quality and quantity. Regeneration may be accomplished by heating a small fraction of the feed water to 80°C. Low grade sensible heat is required and this may sometimes be available as waste heat at no cost. The actual quantity of regenerant required depends on the feed water salinity and type of contactor used. For a 1500 mg/l feed water treated in a simple fixed bed design, the gross heat requirement amounts to 84 MJ/m^3 of water. Where prime energy costs are incurred, 75% of this heat is recoverable from the effluent, so that the net heat requirement is 21 MJ/m^3.

B. Capital and Operating Costs

The pilot plant tests have demonstrated that a thermal regeneration process can be operated in standard fixed bed, countercurrent equipment using resin of standard size, and flow rates comparable to those used in conventional demineralization. From general experience with the scale-up of ion-exchange processes it is evident that the process can be operated in units of 4 m diameter and 2 m bed depth to yield 5000 m^3/day, without further engineering development. Because ion-exchange softening

FIG. 33. A Boby CI, Asahi type moving bed contactor showing the cold adsorption (left) and hot regeneration (right) columns.

plants have been built for flows in excess of 100,000 m^3/day, it should eventually be possible to erect very large, single units so as to minimize capital costs.

A detailed engineering design study has been made for a plant yielding 4500 m^3/day (1.25 million U.S. gallons per day) in an Australian capital city location. Capital and operating costs have been estimated for this example, and are tabulated in Table 5. The estimate excludes any pretreatment other than pH adjustment of the feed water. The design is therefore directly applicable to a soft, anaerobic deep well water. Water from other sources may require pretreatment using conventional technology.

From these estimates it can be seen that a thermal regeneration process already offers a competitive desalting cost on mildly brackish waters. It is inherent in ion-exchange processes that full advantage can be taken of large scale operation, so that water costs will be significantly reduced for desalting modules from 10,000 to 100,000 m^3/day.

C. Pretreatment Requirements

Additional costs will be incurred in many instances through the need for a pretreatment stage. In common with other ion-exchange and related membrane processes it will be necessary to remove suspended solids and organic matter from surface or reclaimed waters by conventional clarification and filtration techniques. Full protection of the resins from organic fouling is recommended until operating experience has been gained on a wide range of waters.

Weakly basic ion-exchange resins are susceptible to oxidation at high temperatures. Many deep well waters contain no oxygen, but shallow well water and surface water sources should be deoxygenated in a pretreatment stage. This is currently accomplished by vacuum deaeration. Where clarification of the water is required, a technique of combining deaeration and coagulation by ferrous ions produced by the electrolysis of iron plates has been developed [48].

TABLE 5

ESTIMATES OF CAPITAL AND OPERATING COST
(PRICES IN AUSTRALIAN DOLLARS)

Plant output	4500 m^3/day
Feed NaCl	1500 mg/l
Product	500 mg/l
Pretreatment	nil
Yield	75%
Plant life	20 years
Basis: plant built in 1973	
Estimates exclude feed water supply, product distribution and effluent disposal	
Plant capital	$370,000
Operating cost, including resin replacement	6 cents/m^3 (22 cents/U.S. 1000 gal.)
Total water cost, including amortization and interest	8 to 10 cents/m^3 (29 to 36 cents/U.S. 1000 gal.)

For waters in which sodium is the predominant cation, it is economically advantageous to soften completely the feed water. This may be accomplished by conventional ion-exchange softening using a strong-acid resin regenerated by salt or by using a carboxylic acid resin in its sodium form. It has been shown that in many instances sufficient salt is available in the effluent from the thermal regeneration plant to regenerate the strong-electrolyte resin. Lime or lime soda softening may be used where clarification is also required. With waters high in both hardness and alkalinity, dealkalization can be an attractive pretreatment, as a reduction in salinity is achieved together with softening and removal of bicarbonate.

VII. FUTURE DEVELOPMENTS

There is room for a considerable improvement in the process and a vigorous research effort in this direction is in progress. The challenge to improve resin performance further remains. Improvements can be made to yield systems of high capacity and faster kinetics.

CSIRO has successfully operated an experimental unit on a treated sewage effluent having a total organic carbon content of about 10 mg/l by using a scavenger resin to prevent the slow fouling of the Sirotherm resin which would otherwise occur. Since the renovation of waste waters will be essential within 10 years to meet water demand, and effluents in many locations around the world are suitable for re-use for industrial purposes except for the high salinity brought about by salt water intrusion into sewers, the ability of desalination processes to cope with organic contaminants is vital.

A pilot plant treating 24 m^3/day of sewage effluent has been installed in Tokyo by Mitsubishi Chemical Industries and ICI Australia to test further the procedure. It is designed to yield a product of 300 mg/l quality from an effluent containing 1000 to 1300 mg/l of dissolved salts.

As shown by the data in Table 4, there is a considerable advantage in operating the process in a continuous manner. The application of the process to municipal water supply and the treatment of effluents will require operation on a large scale, so there is need for improved and simplified contactors which can operate in single units to achieve economies in this direction. This is where self-flocculating micro resins of the magnetic type have an enormous potential, since they provide the advantage of faster kinetics and simpler engineering, as mentioned above.

The most rewarding area for future research is in pretreatment. Improved techniques are under development for oxygen removal, since this is essential for all feed waters except those from deep underground sources, and a simple, inexpensive procedure is mandatory. Depending on the type of water to be treated, the preliminary removal of various ions may be necessary. Again, the development of magnetic resin technology, whether the requirement is one for dealkalization, softening, or the removal of organics seems ideally suited for this purpose. Continuous contactors for the removal of inorganic contaminants are under active development. Of prime importance is the fact that the new systems are very well suited to the operation of water treatment processes on a very large scale, because of the great simplicity in both plant construction and operation.

It is anticipated that the application of magnetic resin technology to thermal regeneration and associated pretreatment processes will simplify their operation considerably, and will reduce capital and operating costs substantially.

ACKNOWLEDGMENTS

The authors are grateful to the editor of the Australian Journal of Chemistry for permission to reproduce Figures 4-6, 8-11, 13, 16, 17, 19 and 21 which are from references 18, 23 and 29, and similarly the Elsevier Scientific Publishing Company for the use of Figures 22, 26, 27, 30 and 33, reproduced from Desalination (references 34 and 48).

REFERENCES

1. A. Rose, R. F. Sweeny, T. B. Hoover and V. N. Schrodt, Saline Water Conversion, Am. Chem. Soc., Advances in Chem. Series No. 27, 1960, p. 50.
2. D. E. Weiss, B. A. Bolto, R. McNeill, A. S. Macpherson, R. Siudak, E. A. Swinton and D. Willis, J. Inst. Eng. Aust., 37, 193 (1965).
3. L. O. Stine, L. C. Hardison and A. G. Lickus, U.S. Patent 3,231,492 (Jan. 25, 1966).
4. H. S. Bloch, U.S. Patent 3,351,549 (Nov. 7, 1967).

5. M. J. Hatch, J. A. Dillon and H. B. Smith, Ind. Eng. Chem., 49, 1812 (1957).
6. R. McNeill and D. E. Weiss, Proc. Fourth Carbon Conf., Pergamon Press, New York, 1960, p. 281.
7. D. E. Weiss and B. A. Bolto, Physics and Chemistry of the Organic Solid State, edited by D. Fox, M. M. Labes and A. Weissberger, Interscience Publishers, New York, Vol. 2, 1965, p. 68.
8. B. A. Bolto, Organic Semiconducting Polymers, edited by J. E. Katon, Marcel Dekker, New York, 1968, p. 199.
9. D. E. Weiss, Proc. Royal Aust. Chem. Inst., 34, 261 (1967).
10. I. M. Klotz and V. H. Stryker, J. Am. Chem. Soc., 82, 5169 (1960).
11. I. J. O'Donnell and E.O.P. Thompson, Aust. J. Biol. Sci., 13, 69 (1960).
12. D. D. Perrin, Dissociation Constants of Organic Bases in Aqueous Solution, Buterworths, London, 1965.
13. D. E. Weiss and B. A. Bolto, Aust. Patent 274,029 (Mar.22,1967).
14. D. E. Weiss and B. A. Bolto, U.S. Patent 3,425,937 (Feb.18,1969).
15. H. A. J. Battaerd, N. V. Blesing, B. A. Bolto, A.F.G. Cope, G. K. Stephens, D. E. Weiss, D. Willis and J. C. Worboys, Effluent and Water Treat. J., 14, 245 (1974).
16. S. D. Hamann, Aust. J. Chem., 24, 1979 (1971).
17. S. D. Hamann, ibid., 24, 2439 (1971).
18. D. E. Weiss, B. A. Bolto, R. McNeill, A. S. Macpherson, R. Siudak, E. A. Swinton and D. Willis, ibid., 19, 561 (1966).
19. H.A.J. Battaerd, U.S. Patent 3,619,394 (Nov. 9, 1971).
20. C. E. Schildknecht, Allyl Compounds and Their Polymers, Interscience, New York, 1973, p. 532.
21. D. Willis, H.A.J. Battaerd, G. A. Lang and D. E. Weiss, U.S. Patent 3,888,928 (June 10, 1975).
22. C. D. McLean, A. K. Ong and D. H. Solomon, J. Macromol. Sci.-Chem., Submitted for publication.
23. B. A. Bolto, R. McNeill, A. S. Macpherson, R. Siudak, D. E. Weiss and D. Willis, Aust. J. Chem., 21, 2703 (1968).
24. D. E. Weiss, B. A. Bolto, R. McNeill, A. S. Macpherson, R. Siudak, E. A. Swinton and D. Willis, ibid., 19, 589 (1966).
25. R. L. Gustafson and J. A. Liro, J. Phys. Chem., 69, 2849 (1965).
26. I. Michaeli and A. Katchalsky, J. Pol. Sci., 23, 683 (1957).
27. S. D. Hamann and C.H.J. Johnson, Aust. J. Chem., 21, 2695 (1968).

28. P. Mueller and P. O. Rudin, Biochem. Biophys. Res. Comm., 26, 398 (1967).
29. D. E. Weiss, B. A. Bolto, R. McNeill, A. S. Macpherson, R. Siudak, E. A. Swinton and D. Willis, Aust. J. Chem., 19, 765 (1966).
30. B. A. Bolto and R. E. Warner, Desalination, 8, 21 (1970).
31. B. A. Bolto, A.S. Macpherson, R. Siudak, R.E. Warner, D.E. Weiss and D. Willis, Ion Exchange in the Process Industries, Soc. Chem. Ind., London 1970, p. 270.
32. B. A. Bolto, R. McNeill, A. S. Macpherson, R. Siudak, E. A. Swinton, R. E. Warner, D. E. Weiss and D. Willis, J. Macromol. Sci.-Chem., A4, 1039 (1970).
33. R. E. Warner, A. M. Kennedy and B. A. Bolto, ibid., A4, 1125 (1970).
34. B. A. Bolto, K. Eppinger, A. S. Macpherson, R. Siudak, D. E. Weiss and D. Willis, Desalination, 13, 269 (1973).
35. F. Helfferich, Ion Exchange, edited by J. Marinsky, Marcel Dekker, New York, Vol. 1, 1966, p. 65.
36. B. A. Bolto, U.S. Patent 3,875,085 (April 1, 1975).
37. G. Adams, P. M. Jones and J. R. Millar, J. Chem. Soc., A, 2543 (1969).
38. R. B. Oke and J. R. Millar, Israel J. Chem., 10, 919 (1972).
39. D. E. Weiss, B. A. Bolto, D. Willis, R. McNeill and D. L. Ford, U.S. Patent 3,560,378 (Feb. 2, 1971).
40. N. V. Blesing, B. A. Bolto, D. L. Ford, R. McNeill, A. S. Macpherson, J. D. Melbourne, F. Mort, R. Siudak, E. A. Swinton, D. E. Weiss and D. Willis, Ion Exchange in the Process Industries, Soc. Chem. Ind., London, 1970, p. 371.
41. B. A. Bolto, D. R. Dixon, R. J. Eldridge, E. A. Swinton, D. E. Weiss, D. Willis, H. A. J. Battaerd and P. H. Young, J. Pol. Sci., Symp. No. 49, 211 (1975).
42. B. A. Bolto, D. R. Dixon, E. A. Swinton and D. E. Weiss, Ion Exchange and Membranes, submitted for publication.
43. D. E. Weiss, B. A. Bolto and D. Willis, U.S. Patent 3,645,922 (April 20, 1972).
44. B. A. Bolto, U.S. Patent 3,808,158 (July 7, 1974).
45. B. A. Bolto, H. A. J. Battaerd and P. G. S. Shah, U.S. Patent 3,839,237 (Oct. 1, 1974).
46. M. B. Jackson, J. Macromol. Sci.-Chem., Submitted for publication.
47. H. A. J. Battaerd, B. A. Bolto and P. G. S. Shah, U.S. Patent Ser. No. 360, 188 (Feb. 27, 1975).

48. H. A. J. Battaerd, N. V. Belsing, B. A. Bolto, A. F. G. Cope, G. K. Stephens, D. E. Weiss, D. Willis and J. C. Worboys, Desalination, 12, 217 (1973).

49. D. E. Weiss, B. A. Bolto, R. McNeill, A. S. Macpherson, R. Siudak, E. A. Swinton, and D. Willis, Aust. J. Chem., 19, 791 (1966).

50. F. Helfferich, Ion-Exchange, McGraw-Hill Book Company, New York, 1962, p.85.

INDEX

A

Absorptivity, 85
Activation energy, 23-25
Affinity series, 180-181
$AlCl_3$, 14
N,N'-alkylenedimethacrylamides, 46
Amberlite IRA-938, 43
Amination, of chloroacetylate, 14
Amphoteric resins (see Resins, amphoteric)

B

b, 8
Boundary film, 5-6

C

C, 8
C_o, 8
C_S, 13
C_{SO}, 10
Carbonaceous combustion, 3
Chemical resistance reaction model, 14-19
Chloroacetylation:
  chemical resistance controlled reaction model for, 14-19
  mathematical models for, 17-19
  particle size in, 16
  rate constants for, 18
  temperature effects in, 15
Chloroacetyl-chloride, 14
Chloromethylation, 5
  homogeneous reaction models for, 19-25
  mathematical model for, 21-25
  step-by-step mechanism, 22
Chloromethyl methyl ether, 19

Clay minerals:
  in natural water systems, 169-171
  structure and exchange properties, 175-179
CMME, 19
Cobalt, reflectance spectra of, 106
Complex sorption, 120-141
Continuous reaction model, 4
Conversion, 9-10
  degree of, 6-7
Copolymer:
  kinetics of, 2-20
  sulfonated, 3
  swelling in, 2
Copper, adsorption of zeolites, 110-114

D

D, 8
D_e, 9-10
Dealkylization, 271-274
Deka-methylenedimethacrylamide, 47
Diffusion:
  controlled sulfonation, 6-14
  exchange rate control, 27
  film, 7-9
  fluxes, 6
  intra-particle, 5
  reacted layer, 9-14
  resin rate, 253-254
  resistance to, 12
Diisopropenylbenzene, 57
N,N-dimethylaminoethylmethacrylamide, 52
N,N-dimethylamino-p-phenylmethacrylamide, 52
1,4-dioxybenzene, 55

DMA, 52
DMDMA, 47
Dowex 50, 49
  copper sorption, 125-128
  swelling capacity, 51
DPB, 57
DVB, 30

E

EDMA, 47
Electron spin resonance, 89
  complex sorption by, 120-121,
    128-131, 143-144
  of copper, 110-114
  of manganese, 101-103
Electrophilic substitution, 2-26
Equilibria, ion exchange natural
    systems, 179-192
  adsorption isotherm for, 180-183
  diagram from titration curves,
    242-246
  double layer treatment, 183
  experimental measurement,
    181-182
  mass action laws applied, 183
  resins, 222-253
Equilibrium swelling (see Swelling,
    equilibrium)
ESR (see Electron spin resonance)
Ethylenedimethacrylamide, 47

F

f, 21
Film thickness, 18
First order reaction, pseudo-uni-
    molecular, 21
Flux, 7-8
Frequency factors, 23-25
Friedel-Crafts, 14, 19

H

Half-exchange, 71
Heterogeneous reactions, 2-26
  continuous reaction model, 4
  Shell-progressive mechanism, 2-4
N,N'-hexamethylenedimethacrylamide,
    47
HMDMA, 47

Homogeneous reaction model, for
    chloromethylation, 19-25
Homopolymerization, of styrene,
    40

I

Illite, 179
Iminodiacetic exchangers,
    146-148
Infrared spectroscopy, 87
  complex sorption by, 121,
    131-136, 144
  of organic exchangers, 116-118
  sample of preparation for,
    84-87
  of zeolites, 91-98
Interpenetrating structures,
    35-37
Ionogenic groups, 52
Ion pairing, 234-237
IR (see Infrared spectroscopy)
Iron ore reduction, 18
  adsorption on zeolites, 103-106
Isoporous resins, 56-59
  catalysts for, 58
  equilibrium swelling, 62-63
  macronet type, 59-73
  osmotic stability of, 60
  structure of, 31

K

k, 18
Kaolinite, 176
KB-4, 47, 49
Kinetics, models, 2-27
  of acid uptake, 26
  chain, 23
  elution, 26
  resins, 253, 286
  temperature effect in chloro-
    acetylation, 15
  temperature effect in sulfona-
    tion, 7
KY-2, 47

M

Macronet resins, 46-73
  chemical stability of, 53

(Macronet resins)
 friable polymer, 53, 55
 isoporous types, 59-73
 permeability, differences, 49
 pore size in, 48
 swelling capacity of, 50
 table of, 54
Macroporous structures, 31, 38, 42-46
Magnetic susceptibility, 89
 of cobalt, 106
Manganese, sorption on zeolites, 101-103
Mass transfer:
 diffusion control in, 6-14
 interphase, 5
Matrix, polymeric, 29-73
 preparation of, 32-73
MCDE, 60
Membrane exchangers, spectra of, 148-152
Metallic oxides, in natural water systems, 171-173
Metal sphere oxidation, 3
Methacrylic acid, 46
Methylene chloride, swelling in, 14
2-Methyl-5-vinylpyridine, 52
Microheterogeneous structure, 42
Monochlorodimethyl ether, 60
Mössbauer effects, 89
 complex sorption by, 123
 iron, 103-106
 monomer studied by, 124-125
MSVP, 52

N

Near infrared spectroscopy (see Infrared spectroscopy)
NIR (see Infrared spectroscopy)
Nitromethane, 6
NMR (see Nuclear magnetic resonance)
Novobiocin, 34
Nuclear magnetic resonance, 89
 of organic exchangers, 116-220

O

Oleum-methylene chloride, 6

Osmotic stability, 40
1,12-(p-oxyphenyl)dodecane, 55
2,2-bis(p-oxyphenyl)propane, 55

P

Particle substitution, degree of, 5
pH, 246-247
Polarographic studies, ion binding, 145
Polymeric matrix (see Matrix, polymeric)
Potentiometric studies, ion binding, 145

R

r, 9
$\bar{R}$, 13
Raman spectroscopy, 87-89
 complex sorption by, 121-122
Rate constant, chloroacetylation, 18
Rate equation, 4-6
Reduction, iron ore, 18-19
Reflective spectra:
 of cobalt, 106-110
 complex sorption by, 123-221
Regeneration, thermal, 221-286
 economics of, 281-283
 engineering aspects of, 274-275
 multistage process, 271-274
 pilot plant for, 275-281
 pretreatment for, 283-284
 single stage process, 269-271
Relative reflectance, 85
Resins:
 amphoteric, 255
 composite, 260-265
 equilibria, 222-253
 kinetics, 253-286
 mixed, 250-253
 "no-matrix" type, 265-269
 optimum pH, 246-247
 particle size in, 256-260
 "plum pudding", 263-264
 porous, 255
 rates of diffusion, 253-254
 salt concentration effects, 246-250

S

S, 37
Sediments, ion exchange in, 173-175
Selectivity, ion exchange natural systems, 179-192
　experimental measurement, 181-182
　expressions, 184-188
　measurements, 193-215
　thermodynamics of, 188-193
Shell-progressive mechanism, 2-4
　applications of, 25-26
　rate equation for, 4-6
Sirotherm, 274
Snake-cage polymers, 256
SPMA, 47
Streptomycin, 47
Styrene copolymer, 32-35
　exchange capacity with, 33-34
　freon non-solvating media, 42-46
　homopolymerization of, 40
　swelling capacity of, 33
Styrene-divinyl benzene copolymer, 30
　formation in presence of solvation, 37-40
　structural interpretation of, 35-37
Sulfonated copolymer:
　concentration profile of, 3
　deviation in reaction rates, 11-12
　diffusion controlled reaction, 6-14
　mass transfer model, 6-14
　temperature effects on kinetics, 7
Sulfonation (see Sulfonated copolymer)
4-Sulfophenylmethacrylamide, 47
Swelling:
　capacity, 37-40
　of copolymer, 2
　equilibrium, 63-65
　in methylene chloride, 14

T

Telogenated polymers, 30, 40-42
　permeability of, 41-42

Tetracyclin, 48
Thermal regeneration (see Regeneration, thermal)
Thermodynamics, ion exchange, 188-193
Titration curves, 227-234
　equilibrium diagrams from, 242-253
　overlap of, 237-242

U

U_N, 37
U_X, 37
Ultraviolet spectroscopy
　adsorption by, 136-138
　sample preparation for, 84-87
UV (see Ultraviolet spectroscopy)

V

Visable spectroscopy
　complex sorption determination by, 121, 138-143
　sample preparation for, 84-87

W

Water systems, natural, 165-215
　ion exchange materials in, 169-173
　ion exchange reactions in, 168-169

X

X, 6
X-ray studies, zeolites, 90

Z

Zeo-Carb-225, 36
Zeolites:
　catalytic properties of, 101-114
　exchange properties of, 101-114
　hydroxyl content of, 98-101
　properties, 89-90
　structure, 91-98
　water concentration, 98-101
$ZnCl_2$, 21